"Making climate change meaningful at the local scale is a precondition for empowering local communities to deal with its consequences on their own terms. This volume illustrates how the concept of clime can help achieve this goal by viewing climate variability and change from a Himalayan multi-species perspective."

Theodore G. Shepherd, *Grantham Professor of Climate Science, University of Reading, UK*

"This volume is a veritable tour de force in the decolonial environmental humanities. Bringing dominant climate science into conversation with every-day human experience, it compellingly centers the affective, multispecies, and situated ways that climate change is encountered and interpreted across Himalayan sites, scales, and subjects. Incisive and nuanced in its analyses, this volume is essential reading for anyone concerned with the changing natures of life, entanglement, and agency in an age of ecological unraveling."

Sophie Chao, *Lecturer in Anthropology, University of Sydney, Australia*

"This is not just a book about the Himalayas; it's a conversation starter, a call to action, and an invitation to see the Himalayan world through a lens of multispecies encounters and shared climatic realities. The groundbreaking chapters delve into the Himalayas, not just as a physical terrain, but as a living, breathing entity shaped by diverse climates and vibrant multispecies interactions."

Arupjyoti Saikia, *Professor of History, Indian Institute of Technology, Guwahati, India*

Himalayan Climes and Multispecies Encounters

Woven together as a text of humanities-based environmental research outcomes, *Himalayan Climes and Multispecies Encounters* hosts a collection of historical and fieldwork-based case studies and conceptual discussions of climate change in the greater Himalayan region.

The collective endeavor of the book is expressed in what the editors characterize as the clime studies of the Himalayan multispecies worlds. Synonymous with place embodied with weather patterns and environmental history, clime is understood as both a recipient of and a contributor to climate change over time. Supported by empirical and historical findings, the chapters showcase climate change as clime change that concurrently entails multispecies encounters, multifaceted cultural processes, and ecologically specific environmental changes in the more-than-human worlds of the Himalayas.

As the case studies complement, enrich, and converse with natural scientific understandings of Himalayan climate change, this book offers students, academics, and the interested public fresh approaches to the interdisciplinary field of climate studies and policy debates on climate change and sustainable development.

Jelle J.P. Wouters is Associate Professor in Anthropology and Sociology at Royal Thimphu College, and Chair of the Himalayan Centre for Environmental Humanities.

Dan Smyer Yü is Kuige Professor of Ethnology at Yunnan University, and a Global Faculty Member of the University of Cologne, Germany.

Routledge Environmental Humanities

The *Routledge Environmental Humanities* series is an original and inspiring venture recognising that today's world agricultural and water crises, ocean pollution and resource depletion, global warming from greenhouse gases, urban sprawl, overpopulation, food insecurity and environmental justice are all *crises of culture*.

The reality of understanding and finding adaptive solutions to our present and future environmental challenges has shifted the epicenter of environmental studies away from an exclusively scientific and technological framework to one that depends on the human-focused disciplines and ideas of the humanities and allied social sciences.

We thus welcome book proposals from all humanities and social sciences disciplines for an inclusive and interdisciplinary series. We favour manuscripts aimed at an international readership and written in a lively and accessible style. The readership comprises scholars and students from the humanities and social sciences and thoughtful readers concerned about the human dimensions of environmental change.

For more information about this series, please visit: www.routledge.com/Routledge-Environmental-Humanities/book-series/REH

Himalayan Climes and Multispecies Encounters

Edited by
Jelle J.P. Wouters and Dan Smyer Yü

First published 2025
by Routledge
4 Park Square, Milton Park, Abingdon, Oxon OX14 4RN

and by Routledge
605 Third Avenue, New York, NY 10158

Routledge is an imprint of the Taylor & Francis Group, an informa business

British Library Cataloguing-in-Publication Data
A catalogue record for this book is available from the British Library

ISBN: 978-1-032-77697-2 (hbk)
ISBN: 978-1-032-77700-9 (pbk)
ISBN: 978-1-003-48439-4 (ebk)

DOI: 10.4324/9781003484394

Typeset in Times New Roman
by Taylor & Francis Books

Contents

List of Figures ix
Acknowledgments xi
List of Contributors xiii

1 Himalayan Climes and Multispecies Encounters: An Introduction 1
JELLE J.P. WOUTERS

2 Paddy Clime: Ecological Indigeneity in the Naga Uplands 30
RODERICK WIJUNAMAI

3 Lakes in Life: Mermaids and Anthropocenic Waters in the Bhutan Highlands 48
JELLE J.P. WOUTERS AND THINLEY DEMA

4 Storied Toponyms in Bhutan: Affective Landscapes, Spiritual Encounters, and Clime Change 71
KINLEY DORJI

5 Climing Everest through Cryo-Visuals 87
JOLYNNA SINANAN

6 Dancing in the Rain: Climing the Monsoon in Pre-Modern Assam 106
RIMA KALITA

7 A Thirsty Himalayas: Rain Clime and Anthropogenic Drought in the Darjeeling Hills 123
SANGAY TAMANG

8 Clim(b)ing Slow-Moving Structures in the Garhwal Himalayas 138
AINSLIE MURRAY

9 The Geopolitics of Riverine Climes in the Eastern Himalayas: The Brahmaputra–Yarlung Tsangpo and the India–China Border Conflict 157
ALEXANDER E. DAVIS

10 Encountering Climate Change: Agential Mountains, Angry Deities, and Anthropocenic Clime in the Bhutan Highlands 174
DEKI YANGZOM AND JELLE J.P. WOUTERS

11 Predatory Climes: Beastly Encounters in the Making of the Sundarbans 197
JASON CONS

12 Afterword: A Himalayan–Andean Conversation 215
KARSTEN PAERREGAARD

Index 224

Figures

3.1 Map of the drainage system and rivers in Bhutan 51
3.2 A *tshoyak* 54
3.3(a) Two lakes in the western Bhutan highlands 59
3.3(b) (Cont.) 60
3.4 Map of lakes in Soe, Thimphu, Bhutan, 2006 66
3.5 Map of the increasing size of Tshomphu Lake 67
5.1 Caspar David Friedrich, *Wanderer above the Sea of Fog*, 1818, oil on canvas, 98.4 × 74.8 cm, Kunsthalle Hamburg, Hamburg, Germany 94
5.2 Trekker overlooking valley on the Everest Base Camp trek 95
5.3 A selection of Instagram images tagged with #everest 97
5.4 Section of the Khumbu Glacier seen from the trek to Gorek Shep 101
6.1 Illustrated folio from *Chitra Bhagavata* depicting a monsoon-scape in Vrindavana with a number of nonhumans 114
6.2 Illustrated folio from *Chitra Bhagavata* depicting a monsoon-scape in Vrindavana with the iconographic depiction of local women in the extreme left corner 114
6.3 Illustrated folio from *Chitra Bhagavata* depicting humans and other-than-humans in a hailstorm in the Braj region 116
6.4 Eighteenth-century illustrated folio from *Anadi Patana* featuring a visual representation of "the Great Deluge" 116
8.1 Char Dham Yatra guide map showing the 39 trekking routes 141
8.2 Two pilgrims being carried in baskets on the path to Yamunotri in 2004 146
8.3 Detail from Char Dham Yatra guide map showing a pilgrim being carried in a basket 147
8.4 Skeletal shacks clinging to the side of the path to Kedarnath on the final day of the 2004 pilgrimage season 149
10.1 View of Jomolhari in Bhutan's western highlands 181
10.2 Crossing the snowline en route to the sacred citadel of Ama Jomo 188
10.3 *Lhakhang* on the shore of the lake below Ama Jomo's palace 189

10.4 Ama Jomo's lake and mountain abode 190
10.5 Twin lakes in the vicinity of Ama Jomo's *phodrang* 192
11.1 Landsat 7 image of the Sundarbans showing divisions between the forest and the land beyond 203
11.2 The "just missed" tiger 204
11.3 Shrimp *ghers* in the delta's siltscape 208

Acknowledgments

Himalayan Climes and Multispecies Encounters is the sequel to *Storying Multipolar Climes of the Himalaya, Andes and Arctic: Anthropocenic Climate and Shapeshifting Watery Lifeworlds* (Routledge Environmental Humanities Series, 2023). Both books are the first outcomes from our initiative for multipolar clime studies with terrestrial, experiential, and affective approaches. The sequel further deepens and enlivens clime studies in the Himalayas specially attuned to the multispecies dimensions of the anthropocenic climate and water.

It is another outcome of the generous support of the Himalayan University Consortium (HUC) led by Dr. Chi Huyen Truong (a.k.a. Shachi). Housed at the International Centre for Integrated Mountain Development (ICIMOD), HUC is a network of over a hundred universities and research institutions based in the Himalayan region and around the world. It has emerged as a strong advocate and practitioner of bridging the earth and climate sciences with the social sciences, humanities, policy-makers, and the public. HUC's commitment to co-developing and co-sustaining the Himalayan environmental humanities through conferences, curricula-building, research, and publication is integral to this vision. Many of this book's contributors are indigenous and early-career scholars whose research activities have been generously supported by mini-grants from HUC. We, editors and authors, are grateful to HUC's vision, network, commitment, and funding. In particular we would like to thank Shachi and her colleagues.

Initiated by the co-leads of the HUC Thematic Working Group on Himalayan Environmental Humanities Dan Smyer Yü (2017–2023) and Jelle J.P. Wouters (2023–), this book project emerged from a week-long online workshop held in 2021 and additional HUC-funded individual fieldworks in 2021–2022. The workshop was co-organized by the Himalayan Centre for Environmental Humanities (HCEH) at Royal Thimphu College (RTC) and partnered with My Climate Risk, a Lighthouse Activity of the World Climate Research Programme (WCRP), the Global South Studies Center at the University of Cologne, and the National Centre for Borderlands Ethnic Studies in Southwest China (NaCBES) at Yunnan University. As with the first book, we continue to be grateful to RTC

President Tshewang Tandin, Dean Shivaraj Bhattarai, and Senior Advisor Samir Patel for facilitating, supporting, and encouraging this project.

We would also like to extend our gratitude to our peer scholars around the world who supported this book. They are Tanka B. Subba, Milinda Banerjee, Michael T. Heneise, Sophie Chao, Arupjyoti Saikia, Ted Shepherd, Sunil Amrith, Pema Choden, Sonam Tenzin, Michael Bollig, Kate Rigby, Clemens Greiner, Michael Kleinod, Thom van Dooren, Emily O'Gorman, Astrid Oberborbeck Andersen, Razzeko Delley, Ambika Aiyadurai, Joy L.K. Pachuau, Malene K. Brandshaug, Kalzang Dorjee Bhutia, Iftekhar Iqbal, Vandana Singh, Ruth Gamble, and Anders Burman.

Contributors

Jason Cons is an Associate Professor of Anthropology at the University of Texas at Austin. His work explores borders, agrarian change, conservation, development, climate change, ferality, and frontiers in South Asia (primarily Bangladesh) and Texas. He is completing a monograph on the politics of future making in the Bengal Delta tentatively titled *Delta Futures: Time, Territory, and Capture on a Climate Frontier.* He is the author of *Sensitive Space: Fragmented Territory at the India–Bangladesh Border* (Global South Asia Series, University of Washington Press, 2016) and the co-editor, with Michael Eilenberg, of *Frontier Assemblages: The Emergent Politics of Resource Frontiers in Asia* (Antipode Book Series, Wiley, 2019). He is an editor of *South Asia: The Journal of South Asian Studies* and part of the editorial collective for *Limn.* His work has appeared in venues such as *Cultural Anthropology, American Ethnologist, Ethnos, Antipode, Journal of Peasant Studies, Political Geography, Ethnography, Modern Asian Studies,* and *Third-World Quarterly.*

Alexander E. Davis is a lecturer in International Relations at the University of Western Australia. He received his Ph.D. in International Studies from the University of Adelaide. He is the author of the monograph *India and the Anglosphere: Race, Identity and Hierarchy in International Relations* (Routledge, 2018). His current research focuses on colonial legacies in India's borderlands, particularly the Indian Ocean and the Himalaya. He is head of the Australian Himalaya Research Network and a foundation member of the New International Histories of South Asia Network.

Thinley Dema is an associate lecturer in the Department of Environmental Management at Royal Thimphu College. She completed her MSc. in Ecology and Environmental Studies at Nālandā University supported by an Indian Ambassador's Scholarship in 2022. She has worked on various research projects focused on environmental, societal, and cultural aspects of Bhutan. Her co-authored publications are *Ecotourism and Social Cohesion: Contrasting Phobjikha and Laya Experiences* (Rig Tshoel, 2019), *Territory, Relationality and the Labour of Deities: Importing Raffestin on the Bhutanese Spiritual Landscape* (Rig Tshoel, 2020), and *Cities by*

Women: Urban Space, Livelihoods and Challenges of Women Street Vendors in Thimphu City (Springer, 2022). Her solo-authored works are *Eco-theological and Economic Perceptions of the Three Brother Mountains (Jampelyang, Chenrizi, and Chana-Dorji) in Bhutan's Haa District* (Routledge, 2021) and *Assessing Vegetation Dynamics in Lingzhi Using Normalized Difference Vegetation Index (NDVI)* (Rig Tshoel, 2023).

Kinley Dorji is a Ph.D. candidate at the College of Natural Resources, Royal University of Bhutan. He studies climate change and migration in eastern Bhutan, which is the least studied region in the country. He was born and brought up in this region before traveling extensively around Bhutan and then stationing himself in Thimphu to begin teaching at Royal Thimphu College. He aspires to become an organically grown native Bhutanese scholar to bridge quantitative environmental studies, conducted at both global and local levels, with qualitative studies of indigenously garnered and socio-culturally curated local environmental philosophy. His contribution to this volume is his first qualitative research output.

Rima Kalita is an assistant professor of Art History and Aesthetics in the Department of Fine Arts and Music at Rajiv Gandhi University, Arunachal Pradesh. Concurrently, she is pursuing her doctoral studies at the Department of Humanities and Social Sciences, Indian Institute of Technology, Guwahati. Her primary research focus centers on exploring environmental imaginations within the pre-nineteenth-century Brahmaputra valley. Her research delves into the pre-modern archive of the eastern Himalayan foothills, encompassing a diverse collection of literary, oral, and visual cultural expressions.

Ainslie Murray is an interdisciplinary artist and academic based in the architecture discipline at the University of New South Wales in Sydney. She trained in Architecture at the University of Adelaide (1993–1998) and was awarded a Ph.D. in Visual Arts by Sydney College of the Arts, University of Sydney, in 2011. Her practice-based research explores intangible aspects of inhabitation and is expressed primarily through film and installation. Her work has been exhibited nationally in Australia and internationally in Canada, China, Denmark, Italy, Japan, New Zealand, and the UK. Besides her creative work, she is the co-editor of *Hand & Mind: Conversations on Architecture and the Built World* (UNSW Press, 2019).

Karsten Paerregaard is Professor Emeritus of Anthropology at the School of Global Studies, Gothenburg University. He has done fieldwork in the Peruvian Andes for 40 years as well as multi-site fieldwork among Peruvian migrants in the United States, Europe, Japan, and South America. His research interests are ecology, water, climate, migration, rituals, and cosmology. His current research is focused on climate-induced migration. His books include *Linking Separate Worlds: Urban Migrants and Rural Lives in Peru* (Berg, 1997), *Peruvians Dispersed: A Global Ethnography of*

Migration (Lexington, 2008), *Return to Sender: The Moral Economy of Peru's Migrant Remittances* (UC Press, 2015), and *Andean Meltdown: A Climate Ethnography of Water, Power and Culture in Peru* (UC Press, 2023). His recent journal articles have appeared in *HAU: Journal of Ethnographic Research* (2020), *Environmental Communication* (2020), *Water International* (2020), *Water Alternatives* (2019), *Climate and Development* (2018), and *WIRE's Water* (2018).

Jolynna Sinanan is a lecturer in Digital Anthropology in the Department of Social Anthropology and the Granada Centre for Visual Anthropology at the University of Manchester. She has conducted extensive fieldwork in Trinidad, Nepal, Australia, and Cambodia and has published widely on digital and data practices, digital visual communication, intergenerational mobilities, work, and gender. Her books include *Social Media in Trinidad* (UCL Press, 2017), *Visualising Facebook* (with Daniel Miller, UCL Press, 2017), and *Digital Media Practices in Households* (with Larissa Hjorth et al., Amsterdam University Press, 2021), and she is co-editor of the *Routledge Companion to Media Anthropology* (2023). Her current research is developing a long-term ethnography of the Everest economy.

Sangay Tamang teaches Sociology at the Department of Humanities and Social Sciences, Indian Institute of Technology (Indian School of Mines) Dhanbad, Jharkhand. He holds a Ph.D. degree from the Department of Humanities and Social Sciences, Indian Institute of Technology, Guwahati, Assam. His broad research interest lies at the intersection of environment, ethnicity, and development, particularly focusing on the eastern Himalayas. He has published research articles in various national and international journals, including *Ethnicities, Economic and Political Weekly, Sociological Bulletin, Indian Anthropologist, Himalaya: The Journal of the Association for Nepal and Himalayan Studies*, and *European Bulletin of Himalayan Research*.

Roderick Wijunamai is a Ph.D. student in the Department of Anthropology, Cornell University, and fellow at the Highland Institute, Kohima (Nagaland). Prior to commencing his Ph.D., he was a lecturer in Sociology and Anthropology at Royal Thimphu College, Bhutan. His research is at the intersection of economic anthropology, political geography, and critical indigenous studies. Some of his authored and co-authored essays have appeared in the *India Forum, The Caravan, The Diplomat, Roadsides*, and *Himal Southasian*.

Jelle J.P. Wouters is Associate Professor of Anthropology and Sociology at Royal Thimphu College, and Chair of the Himalayan Centre for Environmental Humanities. Prior to joining Royal Thimphu College, he taught at Sikkim Central University and was a visiting faculty at Eberhard Karls University of Tübingen, under the Excellence Initiative of the German Research Foundation. He is the author of *In the Shadows of Naga*

Insurgency (Oxford University Press, 2018) and *Nagapolis: A Community Portrait* (Barkweaver Press, 2022), co-author of *Subaltern Studies 2.0: Being against the Capitalocene* (University of Chicago Press, 2022), and co-editor of *The Routledge Handbook of Highland India* (2023), *The Routledge Companion to Northeast India* (2023), and *Storying Multipolar Climes of the Himalaya, Andes and Arctic* (Routledge, 2023).

Deki Yangzom holds a bachelor's degree in Anthropology from Royal Thimphu College, Bhutan. A recipient of the Indian Ambassador's Scholarship, she is currently pursuing her MA in Sociology and Social Anthropology at the Tata Institute of Social Sciences, Guwahati, India. Meanwhile, at the Tarayana Centre for Social Research and Development, she is researching climate change and adaptation, gender equality and gender transformative change, sustainable agriculture, and indigenous knowledge.

Dan Smyer Yü is Kuige Professor of Ethnology at Yunnan University and a global faculty member of the University of Cologne, Germany. He received his Ph.D. in Anthropology from the University of California at Davis in 2006. He currently serves as a member of the Advisory Board of Yale Forum on Religion and Ecology (2018–) and a board member of the Chinese Ethnological Association (2023–). He is the author and editor of numerous publications, such as *Mindscaping the Landscape of Tibet: Place, Memorability, Eco-aesthetics* (De Gruyter, 2015), *Trans-Himalayan Borderlands: Livelihoods, Territorialities, Modernities* (Amsterdam University Press, 2017), and *Environmental Humanities in the New Himalayas: Symbiotic Indigeneity, Commoning, Sustainability* (Routledge, 2021). The Swedish Research Council, the British Academy, the Himalayan University Consortium, the China Social Sciences Foundation, and the German Ministry of Education have funded his research over the last decade. Aside from his academic work, he practices regenerative subsistence farming.

1 Himalayan Climes and Multispecies Encounters

An Introduction

Jelle J.P. Wouters

In 1969, the same year that the US Apollo project turned outer space into an extra-terrestrial frontier in its bilateral rivalry with the Soviet Union, the Soviet scientist Mikhail Budyko and the American scientist Willem Sellers independently made a foundational terrestrial discovery. Baptized as the Budyko–Sellers Climate Model and subsequently developed into the theory of polar amplification, it was found that the earth's high latitudes are warming at least two or three times faster than low latitudes (Walsh and Rackauckas 2015). In a positive feedback loop with runaway planetary climate change, the anthropogenic atmospheric concentration of greenhouse gases creates warming effects that cause polar ice to melt. As sea- and terrestrial ice-sheets liquefy, the ice that remains is younger and thinner and therefore more vulnerable to further melting. When the ice disappears entirely in certain areas, darker ocean and land surfaces absorb more heat from the sun in what is known as Albedo decrease, causing additional temperature rise that accelerates ice melt. This observation by American and Soviet scientists preceded the thawing of the Cold War, which, spurred by an escalation of emission-spewing capitalist designs, accelerated the thawing of the earth's polar regions.

The scientifically measured and modeled "meltdown in the north" (Sturm, Perovich, and Serreze 2003), as well as changing ice conditions in Antarctica, has cascading and profound impacts on natural and human systems in mid- and low-latitudinal regions (Previdi et al. 2021; Esau et al. 2023). Ever since this discovery, and because of global climatic relations and implications, high-latitudinal regions, particularly the warming Arctic Circle, have been at the forefront of climate change research, modeling, funding, theory-making, policy, speculation, and anxiety (Serreze 2018) as well as capitalist exploration (Oberborbeck Andersen 2023) and new geopolitical scrambles (Dodds and Nutall 2015). It was high-latitudinal areas that became seen as the frontline of planetary anthropogenic climate change.

Altitudinal Amplification

Only recently, less than two decades ago, did it first become evident that climate change amplification processes are equally attached to high altitudes,

DOI: 10.4324/9781003484394-1

and especially to the highest of them all: the Himalayas and its adjoining mountain ranges. As a result, the dominant focus on the "melting north" was complemented by a science and policy attunement to the "melting Himalayas" (Xu et al. 2009). Mountaineers, glaciologists, other scientists, and local pastoral communities and other highland dwellers began to report extraordinary and rapid transformations in temperature and glaciation in the remote and largely inaccessible high Himalayas. These transformations interlinked with interruptions and alterations in the local, regional, and planetary climate-regulatory role of the Himalayas, so that experienced changes in hydrological cycles, ecosystems, and natural habitats in lower altitudes and alluvial plains could be explained in reference to temperature change and glacial melt in the cryospheric high Himalayas.

These initial observations were followed up by systematic science-making that corroborated the hypothesis that the Himalayas are warming significantly faster compared to global averages (ICIMOD 2015; 2023), quite possibly at rates that outpace even the Arctic. More than this, the unpredictability of Himalayan climate futures and the dangers of a tipping point and threshold toppings, such as the complete vanishing of glaciers, the disappearance of permafrost, and shifting precipitation patterns, potentially trigger far-reaching runaway climate change, just as the polar regions do.

In global terms, anthropogenic transformations in climate are habitually linked to the "great acceleration" (McNeill and Engelke 2016). This coinage captures the dramatic and consequential post-World War II expansion of human impact on the environment. Other theories offer a deeper time-frame of anthropogenic transformations of the environment that, apart from the agricultural revolution in a general reading, highlight the impact of the "Columbian exchange" and "neo-Europes" in New World countries (Crosby 2004; Smyer Yü 2023, 6–8). Transmitting the discourse of "acceleration," Eriksen (2016) refers to the post-Cold War period as the "acceleration of acceleration," an overheating characterized by a further accelerating surge across wide measures of anthropogenic impacts on the earth. From the vantage of latitudinal and altitudinal highlands, anthropogenic transformations of climate and ecology are further accelerated through a series of cascading effects, teleconnections (see Wouters 2023a), runaway processes, and tipping points. This distinguishes the polar regions and the Himalayas, in earth terms, for accelerating the "acceleration of acceleration."

In the Himalayan contexts, lower altitudinal regions linked to high-altitudinal ecological and climatic affordances are densely inhabited by approximately two billion people, thus firmly distinguishing Himalayan climate change for affecting so many human lives and lifeways so rapidly. The types and ways of other-than-human lives that are affected are many times larger still. These relations between high and low altitudes are so intimate and linear that from geological, ecological, and climatic perspectives, the high Himalayas and the river basins and alluvial plains that surround it are part of the same climate-making and -changing processes. And this matters to the climate sciences, as much as it should imbricate our social sciences and humanities theory-making.

The initial recognition of the Himalayas as disproportionately affected by—and affecting and accelerating—anthropogenic climate change met with an overall scarcity of scientific knowledge about the high Himalayas. As recently as 2007, the Intergovernmental Panel for Climate Change (IPCC 2007, 4) described the Himalayas as a "white spot" with "little to no data." However, the critical appreciation of the Himalayas as a regional and planetary climate-maker and -changer spurred scientific attention to Himalayan climate change. It was in a frame of sudden urgency that the "vast machine" of climate science (Edward 2010) descended on the Himalayas.

However, these scientific efforts were obstructed by problems of physical accessibility, human health hazards, and geopolitical sensitivities and rivalries that reduced climate change research and knowledge to privileged national security interests. Thus, despite augmented scientific efforts, in 2013 the IPCC again admitted that Himalayan climate change and especially its high-cryospheric dynamics and linkages remain "poorly known" and that this "makes determining their future evolution particularly uncertain" (IPCC 2013, 346). Also contributing to this scientific scarcity was the noticeable underrepresentation of Himalayan scientists and specialists in the IPCC, a gap that has gradually been redressed (Alfthan 2017). Since 2013, governments, universities, and other scientific bodies and policy and planning organizations have directed unprecedented resources and attention to understanding Himalayan climate change, with notable results (Xu et al, 2009; Shekhar et al. 2010; Pant et al. 2018; Langenbrunner 2020; see ICIMOD 2023 for an overview). An acclaimed report, published by the International Centre for Integrated Mountain Development in 2023, and with a focus on the cryosphere, summarized the current state of knowledge. It warned:

> The HKH [Hindu-Kush-Himalaya] cryosphere (glaciers, snow, permafrost) is undergoing unprecedented and largely irreversible changes over human timescales, primarily driven by climate change. The impacts are becoming increasingly clear, with increased warming at higher elevations, the accelerated melting of glaciers, increasing permafrost thaw, declining snow cover, and more erratic snowfall patterns. The "water towers" of the HKH, critical for downstream regions, are some of the most vulnerable to these changes in the world.
>
> (ICIMOD 2023: vi)

The report merges hazards, vulnerability, exposure, and risks, and presents the Himalayas as moving beyond the scope of possible climate adaptation and firmly into the domain of loss and damage. Also diagnosed is the coming to an end of the Karakorum anomaly, which refers to the stability or anomalous growth of glaciers in the central Karakorum, in contrast to the retreat of glaciers elsewhere, and especially rapidly so in the eastern Himalayas. The

Karakorum, too, is melting. Policy discussions on the melting Himalayas stress how terrifyingly off-track ongoing interventions are to meet the—already unsafe for the Himalayas—global climate obligation of limiting global warming to within 1.5 degrees Celsius, and speak of a narrow window of opportunity to reverse, or at least stabilize, current trends. This policy talk, as it unfolds in Himalayan conferences and boardrooms, continues to be shot through with concerns about weak inter-state coordination, data deficiency, knowledge gaps, problems of sharing mechanism, and proper data utilization, even though a great deal more is known about Himalayan climate change than it was, say, two decades ago.

In overall terms, the framing of Himalayan climate change by climate science is dominantly organized through numbering practices and modeling that are subsequently translated and enacted through political, bureaucratic, and technical interventions in Himalayan states. These interventions variously take the form of "climate proofing" economic growth, disaster management, or national security paradigms. However, the complex ways in which scientific findings, predictions, and numbers in the "truth-effects" of climate change (Wouters 2023a, 263) come to matter and inform politics, policy, and governance have yet to receive significant scholarship in the Himalayas, the way they have elsewhere (Knox 2020). In broad terms, climate-science-making in the Himalayas remains significantly structured by deep-seated colonial, capitalist, and nationalist systems and, concomitantly, by a range of anthropocentric discourses that readily isolate the human dimension, both in cause and effect, of climate change. While invaluable in scientific terms, this knowledge simultaneously appears as reductionist when considering the Himalayas as embodied Deep Time (see Irvine 2020), indigenous (in the sense of earth architecture and agency) climate-makers and -changers, and further because Himalayan communities are inherently multispecies in their varied constitutions. What is missing, by and large, is an attunement to affect, entanglements, and—in the Himalayan contemporary—disruptions to all types and forms of life, not only humanity.

Whereas scientific knowledge about Himalayan climate change is of recent emergence, Himalayan indigenous ontologies, epistemologies, and genealogies of knowledge and praxis regarding geology, glaciers, water, ecosystems, weather, and climate have much deeper and richer histories locally. Yet, indigenous-vernacular knowledge traditions and lived experiences of geo-ecology and weather–climate occupy only marginal space in formal Himalayan climate change knowledge and policy discussions. Significantly, this is because climate change knowledge production in the Himalayas, as elsewhere, progressively privileges secular–scientific–technical knowledge over other ways of knowing. Chao and Enari (2021, 35), and others with them, note the "exclusionary scope of voices and beings" that are being heeded and represented by dominant climate imageries. It is indeed important to critically assess what forms of knowledge–being–relating climate science displaces, dispossesses, and disavows; what epistemic violence and violent exclusions take place in the assumed superiority of secular, technical, and scientific climate knowledge; to probe, that is, the gap between scientific representations and real-life experiences and understandings of climate change in the Himalayas.

Thus, the representational and epistemic capacities of Himalayan climate science generally fall short in substantially engaging the experiential qualities and meaning-making processes of changing conditions of Himalayan life; of capturing, that is, the terrestrial and felt experiences of those humans and other-than-humans who live the reality behind climate science statistics. Recent ethnographies, including this book's chapters, now show that people across the Himalayas talk about climate, weather, and noticeable changes in the environment, but that they do so in ontological, epistemological, and ethical frames that readily move beyond the expertise and limits of climate science (Mathur 2015; Chakraborty and Sherpa 2021; Gagné 2019; Bhutia 2023; Delley and Aiyadurai 2023). These are accounts produced by and on people who in the face of imposed invisibility in climate change knowledge have something important to say and share about what it means to live in the Himalayas under the sign of the Anthropocene. These are further accounts that foreground the ecological and multispecies relationality of climate change that, across Himalayan geo-ecological niches, are experienced through intricate webs of interconnections and relationalities across and beyond the bios, ecos, atmos, and geos of life.

It is in this context that this book initiates "Himalayan clime studies" and "multispecies," "multibeing," and "more-than-human clime studies" to complement, enrich, and ultimately transform the current scientific race to understand and situate Himalayan climate change with emplaced, lively, experiential, relational, and multispecies accounts of climate change. The Himalayas under investigation here are the indigenous-turned-anthropogenic Himalayas, or the "New Himalayas" (Smyer Yü 2023). Adapted from the "New Arctic" (Evengård, Nymand Larsen, and Paasche 2015; Serreze 2018), the "New Himalayas" refer to the anthropogenically impacted, fragmented, impounded, degraded, and geo-engineered Himalayas, which are the Himalayas of the twenty-first century; the contemporary Himalayas that are affected by, and affecting, local, regional, and planetary climate change.

Himalayan Multispecies Climes

Framed in clime studies (Smyer Yü and Wouters 2023), environmental humanities, and allied knowledge fields, this book considers Himalayan multispecies climes. Building on Fleming (2010) and Carey et al. (2014), we innovate clime as a multispecies dwelling place embodied with weather, climate patterns and change, ecological networks, and geological histories and forces. A clime can be usefully construed as a place where multiple entities and elements combine to produce a more-than-human assemblage whose significance cannot be reduced to these prior entities and elements, but emerges only in relation and co-becoming. Consequently, the whole of a clime is always more than the sum of its parts. Jason Cons (Chapter 11, this volume) writes:

> [T]o think of place from the standpoint of clime, then, is to begin to disaggregate the parts/whole relationships—to see every instantiation of timeplace and every experience of socionature as both produced by and producing broader patterns of environmental and climate change. It is to put mutual constitution back on the table.

Clime thinking therefore foregrounds ecological and multispecies relationality and affirms how varied organisms' practices, adaptations, and experiments are co-making lives and worlds.

Our clime approach partakes of Tim Ingold's (2010, 9) description of "weather-worlds," with weather revealing a relationship with specific material surroundings: "In this mingling, as we live and breathe, the wind, light, and moisture of the sky bind with the substance of the earth in the continual forging of a way through the tangle of life-lines that comprise the land." Expanding this view, clime studies attunes to climate (change), which is more-than-weather and emphasizes uniquely emplaced histories, patterns, and co-becomings that make (and unmake) multispecies life-forms. Clime, as also a word for worlding, is a concept adequate to geo-ecologically nested, complexly connected, dynamic, responsive, historically situated, and affective networks of life in an Anthropocene context. Climate change, as this book's chapters show, necessitates processes of "re-climing," which evokes a resilient and creative but equally a confusing and unsettling process of adaptation and adjustment to changing eco-climatic conditions and the reimagining of self and others (whether human, elemental, vegetal, animal, eco-material, or numinous). These processes of re-climing have now become integral to the contemporary Himalayan condition.

Thus, multispecies clime studies inaugurates a re-situating and rethinking of humans and other-than-humans as geo-eco-climatic beings and relatings. It was first coined and enlivened to climatically interlink the world's altitudinal and latitudinal highlands (specifically, the Himalayas, Andes, and Arctic), which are central to global waters, and so in earth, historical, and ethnographic terms (Smyer Yü 2023). The focus was on the intertwinement of climate, nature, culture, and place; to draw lines and connect dots between changing ecologies, lifeworlds, and livelihoods; and to bring lived and lively climate accounts into conversation with climate science. This cleared the ground for the mutually nourishing inhabitations of clime and climate, which was envisaged as "multilateral clime studies" (Wouters 2023a). While clime remains an open-ended concept that is to be variously filled out and critiqued, it is envisaged, as both a noun and a verb ("climing"—to ecologically, socially, ritually, and spiritually encounter changing landscapes and climates; see Paerregaard 2023), to critically capture the diverse terrestrial embodiments of climate change, with "embodiment" relating affective places as recipients, makers, and changers of climate (Smyer Yü 2023).

This book furthers and expands clime studies with chapters showing how multispecies assemblages and entangled agencies are deeply and diversely entwined with Himalayan climate history and change. Contributions focus on mountains (Chapter 10), tigers (Chapter 11), paddy (Chapter 2), rain (Chapter 7), monsoon (Chapter 6), ice (Chapter 5), earth architecture (Chapter 8), toponyms (Chapter 4), and lakes (Chapter 3). This book will also continue the earlier multipolar clime–climate change conversation through an afterword by Karsten Paerregaard (Chapter 12). A contributor to the first clime book, Paerregaard brings our Himalayan multispecies approaches and findings into dialogue with the Andean worlds he has studied for the past four decades.

In our approach, we collate the terms "multispecies," "multibeing," and "more-than-human," and take these to encompass both different "species of beings," including human, animal, and plant (Chao and Celermajer 2023), and nonsecular/numinous beings (Chapters 3 and 4, this book), as well as a multiplicity of life-forms and life-enabling actants and processes in all their varied epistemological surroundings and temporalities and orogenies (Searle 2013), including geological formations (D'Avignon 2022), glaciers (Gagné 2019), waters (Klaver 2022), plants/plantations (Chao 2022), forests (Kohn 2013), sand-wind (Zee 2022), places (Smyer Yü 2014), territories (Povinelli 2016), and so on.

By allowing the scope, substance, and spectrum of "multispecies" to emerge from Himalayan epistemologies, ethnography, and "lively ethography" (Van Dooren and Bird Rose 2016), our chapters confront colonial–capitalist logics and biological determinism that draw up a straightforward epistemological and ethical boundary between life and nonlife (see Chao and Celermajer 2023). Instead, our contributors are attentive to both the relational nature of earth materialities in the co-creation of meaning and experience, and to Himalayan grammars of agency, animacy, and cosmo-political processes that expand and multiply our understanding of life in the Anthropocene context. Our overall understanding of "multispecies" lies in affirming the relational being and becoming of organismic, geological, ecological, spiritual, and climatic realities. Our explicitly Himalayan focus, in turn, affirmatively recognizes how the Himalayas are local, regional, and planetary makers and changers of climate, as currently experiencing anthropogenic climate change at rates two or three times as fast as the global average, and for the relative dearth of situated, indigenous, and lively accounts that affirm that climate change is always also part of a cultural process, is experienced in specific places, and is lived and adapted to with others, most of whom are not humans.

Such calls for a pluriversality of epistemologies remain outside mainstream knowledge-making of environment and climate in the Himalayas. Ever since the IPCC's "white spot" remark in 2007, the Himalayas, as noted, have witnessed an upwelling of climate science-making. To speak of modern climate (change) knowledge in the Himalayas entails the invocation of scientific specializations and methods that, while internally diverse, are broadly grounded in a set of epistemic convictions about climate, and about how it is possible to gain knowledge of it (Hulme 2009). In its epistemic culture, climate science presumes that any and all aspects of climate (change) can ultimately be measured and modeled, apart from place-specific social and ecological relations and realities, and translated into numbers. There are obvious questions about where these models actually exist on the ground, and where their numbers and invisible quantities reside. What is for certain is that, in their number practices, these models present a logical, often near rhythmic, pattern that seems to function flawlessly in a seemingly random nature. In biology (as in climate science), a model, as Donna Haraway (2017, 228) explains, is a "work object"; a "model is worked, and it does work"; models are "stabilized systems that can be shared among colleagues to investigate questions experimentally and theoretically." A model is not a "metaphor or analogy," nor, indeed, a linear abstraction of felt realities.

Clearly a virtue of this distanced (and disembodied) modeling approach is that all changes in climate, in whatever variables and circumstances they may occur, become reducible to this abstraction (Hulme 2013b), which subsequently relate "truth-effects" in terms of internal coherence and consistency (Wouters 2023a). Models also readily permit for regional and global comparisons. A clear disadvantage of this form of knowledge-making is that numerical, modular, and overall abstract representations do not capture and convey variously emplaced imaginations and experiences of Himalayan communities of life, along with their co-constitutive earth and elemental entanglements. This is because, after the 2007 IPCC report, the Himalayan climate science that formed, in its drive to link the Himalayas with universal properties and propensities of climate variables, significantly alienated the terrestrial and multispecies medium through which climate change is encountered and experienced by Himalayan communities and other-than-human habitats.

While modeling, as a representational practice and mode for prediction, remains the dominant form of scientific climate knowledge production, including in the Himalayas, an emergent strand of scientists are reflecting on the institutional and epistemic limits of climate science. They are reaching out to the social sciences and humanities in an attempt to refashion and expand climate knowledge fields by exploring complementary ways of knowing and representing climate change (Shepherd 2018; Shepherd and Truong 2023; Singh 2023; Hulme 2013a). There is also an albeit more marginal counter-movement of social scientists and humanists substantially engaging with indexical, ordinal, and performative powers of climate science numbers (Amrith 2018; Knox 2020; Zee 2022). Ultimately, the question for the Himalayas, and beyond, is how to mutually complement and also transform the climate sciences, social sciences, humanities, and policy to create more critical and capacious openings for studying and responding to climate change. Clime studies, both multipolar (Smyer Yü and Wouters 2023) and Himalayan and multispecies (this volume), offers itself as a bridge in this collaborative and genuinely transdisciplinary endeavor.

Thus, this book posits that comprehensively conceptualizing Himalayan climate/clime change requires us to move beyond the ontology and representation of the climate sciences to also take into account the diversity of ways of being–knowing–relating in the Anthropocene context. This is not because scientific climate knowledge is necessarily alien to or unsuitable for

Himalayan communities, or indigenous cultures in general; far from it (Kimmerer 2014; Snively and Corsiglia 2001; Tsosie and Claw 2020). It is rather because, in diverse Himalayan epistemologies and cultures, weather and climate (change) are known and experienced within an agential meshwork of matter and meaning within shared communities of life that, in turn, are inextricable from the places in which they are embodied. This understanding of climate change, why it occurs, and the ways in which it is encountered transcends (and transforms) the constitution of what dominant climate science counts as knowledge, and who gets to produce and apply it.

In this spirit, it is to enrich, broaden, and diversify conceptualizations of climate that this book foregrounds views of climate history, pattern, and change from the ground and through everyday experiences in varied geo-ecological niches and gradients. To place climate change in an everyday context, as our chapters do, is not to say that it is a taken-for-granted aspect of local worlds. However, it importantly affirms climate change as recurrently rendering and reorienting everyday experiences in the Himalayas. Furthermore, to situate climate change knowledge in a terrestrial frame accentuates the obvious (yet strangely marginal in most climate sciences) point that humans and most other living beings are "terrestrial creatures" (Ingold 2010), in the sense that they live on the ground, that their knowledge and experience of climate patterns and change are similarly grounded, and that we therefore need to complement often abstract climate knowledge with grounded theorizing of climate (change). With grounded clime/climate change theory we mean the terrestrially attuned, lived, and lively explorations of specific climate "universals" with recourse to how climate pattern and change co-create and encounter human communities and other-than-human habitats, and demonstrably inform everyday experiences, conceptualizations, and representations at different moments and up to the Anthropocenic present.

Thus, whereas climate science in the Himalayas is generally positioned above the ethnographic and multispecies fray, rather than within it, Himalayan clime studies complements, and potentially transforms, this sole focus on scientific approaches with analyses grounded in the experiences and knowledges of Himalayan communities. Their socialities, ethics, and normworlds have always encompassed other-than-human beings, but they are now challenged by unprecedented changes in their environments. In the context of the global climate crisis and planetary unraveling, such a focus on terrestrial experiences allows us to critically apprehend the specificities of changes and reorientations in the very places where they materialize and come to matter.

As our chapters show, from Bhutan to Bangladesh, India to Nepal, changing climatic and ecological conditions are "already-here" and variously reconfigure senses of place, personhood, and time in the Himalayas. However, in addition, they demonstrate that Himalayan communities are the ontologists of their own changing worlds; changes that are readily read and responded to in religious, ritual, cultural, and moral frames. Needless to say, these Himalayan communities of life are diverse, variously constituted through complex historical interactions, and grounded in the specificities of place, time, and community. In its orientation, multispecies clime studies affirms and transmits this diversity of knowledge and experience. In enlivening the experience and idioms through which Himalayan communities experience and articulate climate (change), this book ipso facto pushes back against the categorization of the global climate crisis as a seemingly intractable and incomprehensible "hyper-object" (Morton 2013). Rather, clime studies premises that, for social scientists and humanists, climate change should be regarded as one of many "traditional" topics (along with religion, kinship, culture, and language) that social scientists study to apprehend the institutions, inner logic, and indeterminacies of the communities under investigation.

It is important to note that multispecies clime studies and clime theory do not merely represent a shift from the global to the local, universalities to particularities, abstract representations to felt realities, deductive to inductive reasoning, and material to meaning, but also seek to understand clime–climate (change) with reference to both its planetary determinations and the ways in which it becomes locally lived and responded to. In its intra-actional (Barad 2007), metabolic, and overall relational enactments, a clime is always knotted at many scales (it is the simultaneous presence of the past, the present, and the future), ever in motion (shapeshifting, flowing, moving, eroding, reconstituting through earth spheres and systems), and multiply connected with other entangled agencies, both organic and not. Thus, clime is a deep, profoundly relational concept. In affirming the multispecies mutualities and relationalities that together produce climes, a clime view negotiates and blurs separate spheres (such as ecosphere, geosphere, and hydrosphere) and epistemes (scientific, secular, material, indigenous, spiritual, affective), and adds relational, experiential, and terrestrial perspectives to climate (change) conceptualizations. In this way it expands climate knowledge in the relational imagination, intertwining scientific, technical, and policy issues and concerns with diversely lived and lively experiences, stories, multispecies relations, moral topographies, spiritual geologies, emplaced imaginations, and more.

As the chapters of this book variously illuminate, climes are places that emerge in the co-becoming of social and terrestrial landscapes, in the sense of clime studies foregrounding the terra that everywhere girds experiences of and engagements with climate (change). In other words, terrestriality and climate evince and enact one another in variegated dwelling places, setting in motion climes whose complex geochemistry, physics, metabolism, and assemblages of entangled agencies and relations make and change worlds shared by humans and other-than-humans, and do so at once in material, experiential, somatic, spiritual, and cognitive terms. Clime, then, highlights clime/climate pattern and change as a complex assemblage at the multispecies entanglements of human, animal, plant, ecosystem, and geophysical forces and temporalities. In turn, clime change pertains to a perceivable, experienced, and multidimensional transformation and shapeshifting of the relations between environmental materials, weather conditions, human societies, and other-than-human habitats.

Multispecies Life in the "New Himalayas"

This book maintains that climate change is an inherently multispecies constellation. It affects all species, spheres, and substances of life, as well as the relations between them. However, the dominant scientific–secular–technical ways in which Himalayan climate change is imagined and represented are insufficient to attend to the multiple dimensions and experiences of climate change in terrestrial and multispecies worlds. Specifically, the predominantly human-centric and secular-scientific frameworks that globally drive knowledge production of climate change generally prove inept to meaningfully incorporate the relations, interests, lifeways, and communications of other-than-human beings, and indeed many of the human members of Himalayan communities. To expand and multiply climate knowledge, multispecies clime studies ask how anthropogenic climate change is experienced and encountered by multispecies communities of life; how it reconfigures the relations of humans to each other, to other species, biotic and not, and to the geo-ecological materialities that surround and emplace them.

In order to make us think about climate change as a multispecies affair, Roderick Wijunamai (Chapter 2) invites us to the Naga uplands in India's Northeast. Grounded in ethnography and his own embodied experience as an indigenous Naga scholar, Wijunamai enlivens the multispecies and multidimensional relations of paddy in the making and unmaking of more-than-human worlds. He proposes the notion of "paddy clime" to affectively affirm the centrality of paddy in Naga histories and, up to the unfolding present, its integrality to local ecosystems, soil, environmental flows, multispecies worlds, cultures, and Nagas' affective and spiritual relations with the fecundity of paddy fields. Wijunamai shows how "paddy and humans partake in an ongoing multispecies story; a story that has mutually transformed paddy and humans." Among the Naga, anthropogenic transformations, including climate change but also the bearings of the capitalist market, are limiting the availability and fecundity of soil, precipitation patterns, the production of local crops—especially paddy—and with that the food, bodies, and well-being of Naga villagers. Many Nagas are now compelled to consume non-local varieties of rice that are imported, instead of the much relished "indigenous rice," and this has repercussions that are at once physical, nutritional, emotional, cultural, and spiritual. Among the Naga, as Chapter 2 relates, climate change is significantly engaged and experienced in terms of changing human–paddy relations.

Anthropogenically induced phenological changes not only pertain to plants, such as paddy, but also manifold animals, including the tigers of the Sundarbans, as elucidated in a lively ethography by Jason Cons (Chapter 11). In the delta landscape of the Sundarbans, tigers are "clime-makers." As apex predators, they perform a critical role in balancing fauna and flora in the region, allowing mangrove trees to flourish—"to become the Sundarbans." The tiger's status as a clime-maker is affectively recognized by the human inhabitants for whom it is not just a predator to be feared, but also a deity and a symbol of sovereign powers. In its charismatic form, the "cosmopolitan tiger," a term Cons borrows from Annu Jalais (2008) and contrasts with the actual Sundarban tiger, the animal has also become "an icon of national and international conservation, and a sentinel of global climate change." Tigers in the Sundarbans are today variously threatened, resulting in shifting tiger behavior and patterns of predation, pushing them out of the mangroves and into settled communities.

Multispecies clime studies also re-embeds the numinous/spiritual relations that locally co-constitute climate-worlds. Throughout the Himalayas, across religious worldviews, and in diverse ways, mountaintops and glaciers reveal interagentive and animated materialities that are variously venerated as "deity citadels" (Allison 2019), *bla-ri*, "soul mountains" (Smyer Yü 2014), *gnas ri-s*, abode mountains (Huber 1999), or, in Hinduism, as Parvati, Mountain Lady, and the daughter of Himavat, the embodied personification of the Himalaya. Thus affectively, practically, and politically enacted, mountains and glaciers are integral to variously constituted multibeing "norm-worlds" (Banerjee and Wouters 2022) and known to be particularly sensitive to human (in)actions, and attentive and responsive to changing moral, ethical, and ritual commensalities (Gagné 2019). In their relationality, sociality, and mutuality with humans, and as bound to other life-forms and forms-of-life, Himalayan mountains and glaciers generally enact as living materialities that are integral to cosmo-political processes that thus escalate their ongoingly experienced transformations as also a cultural, ritual, and religious problem (see Haberman 2021; Wouters 2023b).

As a chapter in point: Yangzom and Wouters (Chapter 10) trace the agential entanglements between Bhutanese herders and mountains. These mountains are the sacred abodes of *yul lha*—territorial deities who are known to communicate through weather changes. Yangzom and Wouters innovate the affective notion of "mountain clime" to apprehend how mountains have long been vital to highland human life, both for their life-enabling and sustaining glacio-hydrological and geo-ecological affordances and because of their Bon–Buddhist heritage and significance. Herders and their companion species, predominantly yaks, have long arranged their livelihoods, lifeworlds, sense of morality, and everyday ethics around *yul lha* mountains. In this way, mountains in Bhutan emerge at once as "something" and "somebody." Through ethnography and experience narratives, Yangzom and Wouters show how local climing practices—in terms of how and what highlanders know, feel, and sense about weather, climate, and geo-ecological relations—and anthropogenic alterations therein, are reflected and refracted by their imagined and ritually–spiritually mediated relationship with mountains. In the unfolding Anthropocene, mountains, and their autochthonous, nonsecular owners, are changing their behavior and form, including by shedding their ice and snow. Herders are experiencing a "new heat," which they describe as *dhu* ("poison"), or the "poison of the heat" that assaults the snowline. And, whereas in the past, rituals usually had their desired effects in securing the well-being of herders and cattle, highland ways of being–relating–knowing are today challenged by a recurrent failure of ritual. In this multispecies assemblage, climate change significantly manifests as ritual uncertainty. To the highlanders, this is communicative of an experienced breakdown in the mutualism between mountains and humans. But while these changes are unprecedented, they are accounted for in a prophecy that speaks of seven suns combining, and this heralding *zamling kabenob*—the end of the world.

Such emplaced, affective, and multispecies frames of indigenous knowledge and experience seldom figure in Himalayan policy discussions, which are at once distinctly anthropocentric and secular-scientific in orientation. Kinley Dorji (Chapter 4) laments this omitting of indigenous wisdom traditions and practices, and confesses that his scientific-technical training in environmental and climate science, while evidently enriching, distanced him from both his upbringing in rural Bhutan and his Bhutanese students, for whom "the environment was sentient and storied" and replete with *mi mayin* ("other-than-humans") in shared social worlds—perspectives that are generally erased in scientific approaches. He reflects that the syllabi, assignments, and examinations that guide the teaching of environment in Bhutan "did not allow for our emplaced and embodied knowledge to be considered seriously," which made him "more acutely aware of a wedge between the environments my students and I 'lived in' and the environment I was made to teach." Across the Bhutan landscape, places are stories, and stories are places. These stories, Kinley Dorji explains, intersect and interact with one another in a dense network of intertwined histories, spiritual encounters, cultural memories, and moral frames. Here geology manifests as a many-layered earth text, which the Bhutanese must read and heed in order to secure their overall well-being and spiritual growth. Calling for the complementarity and translatability of science and indigenous knowledge, Kinley Dorji writes:

> It was my return to these stories of my upbringing that made me realize that while scientific–technical knowledge offers a privileged window … science-only knowledge was not doing sufficient justice to the diverse, lively, and affective experiences and knowledge traditions that exist in Bhutan.

The storying-climes approach he presents in his chapter allowed him to draw lines and connect dots between climate science and indigenous knowledge.

The Relational Himalayas

The history of the "indigenous earth" (Smyer Yü 2023, 14), in the sense of a Deep Time, pre-human earth, is largely a history without the Himalayas. Many geological pasts and climates preceded the present (Yusoff 2013), and in terms of geological matter, time, and agency, and as also evidenced by paleoclimatology, for the longest time the earth has been more horizontal, flatter, and hotter than it is today. It was the merger—through continental drift, a theory coined by Alfred Wegener in 1912—of the Indian and Eurasian tectonic plates, beginning "only" some fifty million years ago, that reopened the earth's crust to complement the already existing Andes and Alps and add new geological and ecological—and later social and scientific (Fleetwood 2022)—verticality to the world.

The evolution of Himalayan orogenies (in the plural, as most geologists highlight the diverse and multiple geological histories of the Himalayas) was long and complex, and has many geological chapters and chambers (Searle 2013; Pandit 2017). And while the Himalayas may strike a pose of stability and permanence from a human perspective, a Deep Time geological diary would show Himalayan landscapes continuously changing and shifting, with valleys moving, glaciers and rivers migrating, some peaks disappearing and other peaks emerging. These are the geological tales of many pasts. There are also tales of yet-to-comes as Himalayan mountain-making is a geological work in progress. In Chapter 8, Ainslie Murray refers to this geological work as "slow-moving earth architecture" (traversed by "slow-moving pilgrims" who spiritually approach the earth's geological agency and affordances). At present, the Himalayas are simultaneously growing and being wasted away—through erosion, weathering, and collapse. Over the long geological term—tens or even hundreds of thousands of years into the future—levels of erosion will outpace rates of orogenesis, from that moment the Himalayas will be gradually ground down to silt and disappear into the ocean, and the larger Asian region, and the earth, will be flatter once again.

In their formation, the Himalayas both co-created and became central to the Cenozoic geological era, which emerged out of the Mesozoic era (252 million–66 million BCE) and encompasses the past 65 million years—a mere 1.5 percent of the earth's biography. Cenozoic means "recent life" or "new life." The Himalayas are the makings and makers—and increasingly, in the Anthropocene, the unmakers—of this recent and new life, as understood in climatic, ecological, and hydrological terms. This new life, in the sense of changing earth and climatic conditions, enabled the emergence of hominids who evolved into anatomically modern humans through the Cenozoic epochs of the Miocene, Pliocene, Pleistocene, and finally Holocene, which began about 12,000 years ago. Affirming the emergence of the Himalayas as a planetary change-maker, Maharaj Pandit (2017, 49) writes: "The world before and after the birth of the Himalaya was remarkably different," not only physically but also in climatic terms. The Himalayas altered the climatic profile of Asia, variously in its cooling effects, in the formation of 15,000-odd glaciers that appropriated over 10 percent of global fresh water, and by co-creating new meteorological and ecological conditions; and all of these with planetary ripple effects. Once these mountains reached significant elevations, and through their wide range of geo-physical, geo-chemical, and geo-meteorological actions, adaptations, and movements, the Himalayas cultivated vitalities, materialities, and movements that came to prefigure, enable, shape, and sustain the habitability of places, biomes, and ecosystems and their multispecies worlds, and human affective and spiritual relations with their physical environment. It is in this way that the Himalayas are makers of lives and worlds.

From the vantage of clime–climate studies, the Himalayas are neither given nor fixed; nor are they the mere end result of geological processes. Rather, the Himalayas are produced and productive, generated and generative; they are an agential and relational force whose geological, ecological, and climatic agency is structured through its material relationships, ecologies, assemblages, and processes. In tectonic terms, the Himalayas are the material outcome of the coming together and merger of the Indian and Eurasian plates. The resultant geological process of mountain-making is often framed in destructive terms as a collision or tectonic violence. However, it is more productive, in climatic terms, to think of Himalayan orogenies as the life-enabling (in the sense, that is, of recent and new life) material substance of an intimate intercontinental relationship. *Pace* Pandit (2017), the Himalayan geological histories have made multispecies livability possible.

Significantly, because of their affordances, or "ecological services," the Himalayas' geological architecture and agency have been variously sanctified in spiritual terms as the abode of gods and deities across several religions. In Chapter 8, Murray apprehends pilgrimage (so common across the Himalayas) as an act of "spiritual clim(b)ing" in which geology is approached ritually (see also D'Avignon 2022). Specifically focusing on the Garhwal Himalayas, Murray relates spiritual clim(b)ing—the simultaneous physical and spiritual climing of geo-ecological earth architecture—to pilgrims' affective and spiritual interactions with the earth's life-enabling rivers and mountains. In Garhwal, spiritual clim(b)ing is part of a long anthropogenic history, as it is across the Himalayas, whose great mountain halls have been explored for millennia by sages, monks, and others seeking quiet solace (Wouters and Heneise 2023, 24). In the current Anthropocene, spiritual clim(b)ing in Garhwal encounters a paradox. Whereas this act was once reserved for the few, today hordes of spiritual practitioners and tourists annually ascend the mountains, with the result that the geo-ecological structures and earth affordances that are being clim(b)ed for their sacredness are being anthropogenically affected in the process.

The pilgrimage route through Garhwal, as clim(b)ed by Murray, ascends alongside the Yamuna, Bhagirathi, Mandakini, and Alaknanda rivers to the four terrestrial sources of the Ganges—the most sacred waters in Hinduism and life-giving to millions of inhabitants downstream. More broadly, the Himalayas became a vital node—at once as a storage, shapeshifter, and regulator—in the regional and global water cycle whose composition and circulation it changed. Himalayan clime studies affectively affirms the Himalayas' status as, variously, a node, crossing, connector, confluence, and co-maker of much larger hydrological cycles, rather than as an isolated and self-serving actant or end point. The integral relationality of the Himalayas also manifests in view of the circulation of matter (soil and sediments), of which tons are moved down annually. Significantly, this movement takes place through rivers. While many of these rivers seem to source from the Himalayas and adjoining plateaus, from a clime perspective, these waters travel through land and air concurrently, and the water cycle threads together mountains, plains, and oceans in clime–climate-making processes.

In point of riverine fact, some of the largest "Himalayan" rivers, such as the Indus, the Sutley, and the Tsangpo (Searle 2013, 548), which becomes the Brahmaputra (Saikia 2019), were antecedent to the Himalayan mountain uplift, but reconfigured in their shape, size, and siltation patterns as they struck up material relations with that uplift. Through these rivers and erosion, the Himalayas became movers of sand, sediment, and silt that make and nourish flood-plain ecologies (Saikia 2019)—including the Bengal Delta, which is ultimately a Himalayan creation—and interweaving flows of multispecies life. Willem van Schendel (2009, 3; original emphasis) writes, "Bangladesh *is* the Himalayas, flattened out," in the sense that its delta was built by particles of soil transported and deposited there by "Himalayan" rivers. In Chapter 11 of this volume, Jason Cons affirms the Sundarbans as "the obverse of the Himalayas," co-created by the hydrological cycle that ties the mountains to flood plains to the sea, as much as by the long transfer of sediment from the Himalayas to the Bay of Bengal, resulting in the "slow alluvial rendering of the alpine vertical as the delta horizontal." Not only contemporary inhabitants of the Sundarbans but well over two billion people depend on various Himalayan "gifts" directly in agriculture, drinking water, hydropower, and all kinds of livelihoods, and many more indirectly. The ecosystems and other-than-human lives that are variously sustained through Himalayan relations are many times larger still. Put this way, the Himalayas are present in manifold life-forms and flows of everyday life in places that are far removed from the physical mountaintops. This necessitates that we move beyond the limits of conventional mountaintop understandings of Himalayan geography in anthropogenic terms, and instead privilege ecological, climatic, terrestrial, and relational approaches to appreciate the life-enabling and climate co-making features of the *living* Himalayas.

Scenes of Himalayan relationality also partake of atmospheric entailments, with earth elevation acting as a relational materiality to the moisture-laden air carried by Indian Ocean winds over the Indian subcontinent. This results in torrential rains as the clouds are made to rise, condense, and unleash their moisture in the form of rain (Amrith and Smyer Yü 2023). The Himalayas are thus co-makers of the Indian monsoon, the liquid lifeline of Indian civilization, from past to present. In the context of pre-modern Assam, Rima Kalita (Chapter 6) climes a monsoon story of multispecies life, which she enlivens with accounts of humans, rhinos, elephants, transmigrating fish, and plants/crops. The centrality of the monsoon to Assam's multispecies life historically expressed itself through visual, poetic, and historical productions that relate the monsoon as an active protagonist in the making of the pre-modern Brahmaputra Valley in eco-social, economic, political, and multispecies terms—"from military strategy to experimental cropping patterns, monsoonal rains led the march of creating this clime." Kalita concludes: "the convergence of social expressions and environmental events frequently met at one vantage point by that single protagonist—the monsoon."

Because they co-make the monsoon, the Himalayas' southern slopes are heavily forested and highly eroded, whereas the northern slopes and Tibetan Plateau are generally dry and arid as (most) clouds are unable to cross the mountains. This renders the Himalayas as one of the earth's "most sudden and spectacular climatic-geomorphic divides" (Searle 2013, 273). What Searle frames as "divides" is perhaps better captured as inversely relational—as a difference that is not substantive but relational—with the Himalayas actively co-crafting diverse ecologies and biodiversity that allow for multiple multispecies adaptation and flourishing. Himalayan "rain climes," however, are increasingly changing shape (see Chapters 2 and 7, this volume).

Considering this role of the Himalayas in shaping terrestrial–atmospheric relationalities, it is quite plausible to think of the atmospheric Himalayas (see Zee 2022), even if these atmospheric Himalayas are always terrestrially experienced by humans. While much more is to be said of the Himalayas as maker and changer of climate and ecology, for now, suffice it to note that, in broad terms, air circulation and pressure, water, temperature, and flows of sand and sediments are all mediated by and threaded through the Himalayas. And note, too, that these processes interlink the mountains with regional fluvial, pluvial, atmospheric, aquifer, and tectonic and telluric relations that vastly outstretch the geography of the physical Himalayas. What thus makes the Himalayas vitalist and pivotal in climatic terms is that geology and elevation became inherent to, and co-constitutive of, life-giving and sustaining hydrological, ecological, and other earth systems and spheres across multiple nodes, geographical scales, and multispecies ecological gradients.

When the Himalayas are projected in relational terms, it becomes difficult, or even impossible, to pinpoint exactly where the Himalayas begin and end in climatic and ecological terms. This is in contradistinction to readings of geology that are truncated by geographical and geopolitical interpretations such as the Indian, Chinese, and Nepali Himalayas. In earth terms, there is no compelling argument to determine clearly where to divide the main Himalayan ranges from adjoining and adjacent plateaus, ridges, ranges, belts, spines, and spurs, whether terrestrially, tectonically, ecologically, or climatically. Nor is it helpful to separate high- from low-altitudinal areas as these conjoin in the climate-making processes. For this reason, in this book we invoke the Himalayas in their broadest possible sense, even as our chapters' foci lie with the central and particularly the eastern Himalayas, where climes are always co-created by relations and processes that involve vastly larger regions. This means that we desist the perspective of the cylindrical Mercator projection of the Himalayas as the crown of India. Nor is our episteme that of anthropogenic territorialities in which the Himalayas are disrupted, fragmented, and disputed by borders, nations, geopolitics, and academic area studies. Nor, indeed, are we primarily concerned with the Himalayas in their charismatic mountaintop physicality. Instead, we consider the Himalayas in their atmospheric, hydrological, and other dimensions within earth and climatic spheres and systems.

If very broadly sketching the contours of the Himalayas in anthropogenic terms is nevertheless desirable for purposes of geographical comprehension, our overall understanding is attuned to satellite imagery of elevation when adjusted for the curvature of the earth. In this view, the Himalaya Range, in all its elevational extensions, outgrowths, and connections, appears like the spinal cord of Asia. The central Himalayas, including Mount Everest, the tallest of them all, appear as the thoracic from which interlocking vertebrae extend west as the Karakoram Range and Pamir Knot (the main concentration of the world's highest mountains), east in southwest China, northeast India, Burma, and Bangladesh (with extensions into Thailand, Laos, Cambodia, and Vietnam) before dropping into the Indian Ocean, and north via the Tibetan Plateau and Tian Shan into the Altai mountains and glaciers before folding and unfolding into Siberian permafrost and ultimately the Arctic Ocean. If we also attach the river basins and alluvial plains that everywhere link up to the mountains and are directly co-shaped and nourished by them, what comes into purview is a plurality of climes–climates that are intimately associated with Himalayan elevations. These climes–climates, while taking on distinct geo-ecological patterns, are everywhere interconnected, mutually co-shaping each other. Such an expanded vision of the Himalayas permits a holistic view of how Asian earth elevations co-regulate local and regional climates with other atmospheric and environmental flows.

For one thing, it allows us to think of the Himalayas as a terrestrial and atmospheric bridge between Arctic and Indian waters and affirms new scientific climate discoveries that tell of their teleconnections (Kumar et al. 2023; Sundaram and Holland 2022). But not just a connector and circulator of global waters, the Himalayas themselves can be productively thought of as an oceanic extension, a terrestrial ocean, or an "aquatic land" (Smyer Yü 2021, 9), in the sense that they are fully saturated with water, whether at surface, sub-surface, or atmospheric level. This equally applies to the Himalayan flatlands, such as the Sundarbans, the world's largest remaining mangrove forest, where human and other-than-human livelihoods depend on "navigating the mutable boundaries between land and water" (Chapter 11, this volume). The Himalayas, moreover, enjoin water in its multiple materialities, and altitudinal affordances cause waters' liquid and vapory forms to shapeshift into solid form, as glaciers and snow. Himalayan glaciers store the largest amount of global fresh water apart from the polar regions. In their circulation and flow—through melt, precipitation, watershed dynamics, and rivers—these waters converge biotic and abiotic worlds into multispecies communities of life in the mountains, the foothills, and downstream areas. The affirmative recognition of the Himalayas as the globe's "Third Pole" from the 1950s onward—in terms of the earth's terrestrial waters—firmly associated them with the Arctic and Antarctica (Smyer Yü 2023). This tri-polar world has since been expanded to a multipolar world of latitudinal and altitudinal highlands in geological, climatic, hydrological, and ethnographic terms (Smyer Yü and Wouters 2023). Within this multipolar world, the Himalayas, from their sub-surface aquifers and mountain halls to the air above, direct and drive far-reaching and complex intra-actions of atmospheric, cryospheric, hydrological, geological, and ecological processes that significantly bear on the earth's biodiversity, climate, and freshwater cycles.

These connections bypass anthropogenic imaginations and call for conceptualizations that are better attuned to the Himalayas' role, relevance, and relatings to earth and climatic dynamics. As noted, to affirm the Himalayas as both a regional–global hydraulic node and a water-bridge partakes in new emergences in scientific knowledge of teleconnections, feedback loops, and tipping points that reveal intimate linkages between high-latitudinal and high-altitudinal climate variabilities. This visualization allows a thinking of the Himalayas not as a geographical barrier and divide—as it is predominantly construed in anthropogenic terms—but as a connector of the Arctic and Indian oceans, as seen in the Tibetan Plateau's far-reaching impacts on the Arctic (and Antarctic) climate (Wang et al. 2023), in links between variabilities in Arctic ice and variabilities in Indian monsoon patterns, including extreme rainfall (Chatterjee et al. 2022), and in the warming of Indian oceanic and atmospheric waters leading to Arctic oscillation (Jeong et al. 2022); or of the Himalayas as interacting and guiding Indian waters after they shapeshift to vaporous form and move through the atmosphere (Amrith and Smyer Yü 2023); or of the permafrost in Mongolia and Siberia, co-created by Himalayan cooling effects, as co-regulating Arctic ecosystems (Walther et al. 2017). It also allows for thinking of the retreat of Himalayan glaciers as leading to a lower albedo and the release of black carbon, producing new heat that finds its way to the Arctic, where it accelerates the melt of terrestrial and sea ice (Ma et al. 2019); or, more generally, of the Himalayas as an "intercontinental biological highway" (Pandit 2017), an ecological contact zone, a geo-biological bridge, a subterranean hallway, a waterway, or a material conveyor-belt aiding the exchange of biotas, microbes, and matter between high-, mid-, and low-altitudinal regions. These are just a few of the many positive climate and ecological connections and feedback loops that scientists are now beginning to uncover, all of which must have consequences for how we think and represent space, and the role of the Himalayas within it, in the social sciences and humanities.

Under the duress of the Anthropocene, the dynamic composite substance, spatial properties, and relationalities of these intra-actions are shifting, increasingly often, so that the effect from a human vantage point is water in all the wrong phases (liquid, solid, vaporous, fresh, saline) and places. These waters—and the exchanges and relations between them—are now being affected by, but equally affecting, local, regional, and planetary changes in climate. In their central role in the global water cycle, the Himalayas, as noted, are also distinctively revealed as innate weather- and climate-makers. They are actants in an earth–atmospheric architecture of relations that choreograph intimate dances of air, water, temperature, materialities, and wind.

While the biophysical sciences have long critically appreciated the Himalayas as an enabler and sustainer of multispecies communities both high in the mountains and in adjoining lowlands (Shroder 1993; Searle 2013; Pandit 2017; Alter 2019; Kumari, Srivastava, and Chandra Dumka 2021), the social sciences and humanities have been slow to catch up with the geological, climatic, ecological, and affective affordances and connections. Himalayan clime studies seeks to redress this lacuna.

Anthropocenic Himalayan Waters

As several of our contributors as well as scholars in the emergent hydrohumanities (De Wolf, Faletti, López-Calvo 2021) illuminate, it is the mutually constitutive relations between water (in its various bodies, flows, and phases—liquid, vaporous, solid, saline, and fresh), species (including humans), elements, materialities, and technology that are shifting ground particularly rapidly in the New Himalayas, and in the process transforming and reorienting multispecies communities. Himalayan waters' unfolding story is that of their transformation from indigenous, in terms of forms and flows, to Anthropocenic status. Across temporal–spatial settings, this transformation variously results in excess and scarcity of water. It is deeply implicated in anthropogenic climate change and caught up in its effects, more or less directly. The centrality of water in the Himalayas, and its changing status and form in the New Himalayas, is a central theme of this book and is enlivened in affective coinages of "rain clime," "monsoon clime," "ice clime," "riverine clime," and "lake clime."

In Chapter 7, Sangay Tamang presents rain as a protagonist of place-making, ecological knowledge, and cultural traditions, showing, in sum, how the history of Darjeeling is significantly a history of rain. However, rains and waters in the region's hills and mountains have been anthropogenically altered, transforming them from indigenous (in both earth and human terms) to anthropogenic rain climes mediated by colonial and later postcolonial conditions and interventions that altered, appropriated, and channeled the flow of water through the history of Darjeeling. From a land deeply enmeshed in indigenous rain, where water was abundantly available in atmospheric, aquifer, and terrestrial spheres, Tamang traces how anthropogenic transformations have made the hills "thirsty" for the first time. The "thirsty Himalayas" is not just an apt metaphor of anthropogenic impacts on the socio-hydrological cycle, but communicative of a feared future scenario of a Himalayas without glaciers and snow.

The land is also increasingly thirsty on the Ladakh Plateau, high in the Indian Himalayas. The region lies in the Himalayan rain shadow, so precipitation has always been scarce, but life has long been sustained by meltwater from snow and glaciers. Now, though, in the Anthropocene, the snow is increasingly fickle and melts well before the start of the agricultural season. In turn, the glaciers are shrinking and retreating higher up the mountains, so there is less meltwater from them too, and the greater distance it has to travel means it arrives too late to irrigate the fields. This Anthropocenic displacement and delay of the high-Himalayan waters have resulted in local scarcities that are existential. In response, different human, natural, and technical elements have been drawn together in a more-than-human, more-than-natural, and more-than-technical experiment that connects old glaciers, winter melt, gravity, pipes, labor, (inter-)national financing, cold night-time temperatures, drip irrigation, and human ingenuity to create artificial glaciers in the form of vertical ice cones. Winter snow melt is collected at higher altitudes and channeled down to the villages through pipes. When the water finally flows out of the pipes and hits the sub-zero temperatures, it instantly shapeshifts by freezing. What is now ice quickly amasses and forms the cones that store the winter melt. Some of these cones can reach over fifty meters into the sky. As the ice melts in the spring, the water is collected and channeled to the fields through a sophisticated drip irrigation system (Clouse 2019). The human-facilitated ice structures are celebrated locally as ice stupas because of their physical resemblance to Buddhist shrines, but perhaps also because they contain an increasingly precious and scarce "relic" on the Ladakh Plateau: fresh water. Prayer flags are draped around the ice stupas and move with the moods of the wind; they are not only signs of reverence but also provide some shade and a windbreak to slow the melt. In Anthropocenic Buddhism—a religion that is variously confronted, attuned, and adapted to changing climatic realities—the ice stupas are consecrated by the mantras and chants of Rinpoches and Lamas, while lay Buddhists prostrate themselves before them. Geologically speaking, whereas natural glaciers are massive, moving ice sheets, the relatively tiny, static ice cones have few implications for the greater Ladakh region and the Himalayas themselves. Yet, politically speaking, they have been subject to inter-communal tension between upstream and downstream communities because the former divert water to create them (Parvaiz 2018). To be sure, in the past, Ladhakis would nourish glaciers by insulating the ice with charcoal. Back then, it was a community responsibility "to bring charcoal to the mountains and throw it on the glaciers to make it grow" (Gagné 2019, 5). However, the artificial glaciers are an anthropogenic and Anthropocenic indicator and a direct response to regional and local geo-atmospheric permutations that confront the possibility of vegetal and human life.

That waters readily take on sacred and spiritual dimensions, and that anthropogenic alterations therefore also carry ritual and religious implications, is enlivened by Wouters and Dema in their exploration of Bhutan's highland lakes (Chapter 3). These lakes are inhabited and owned by numinous *tshomen* (mermaids) who are deeply invested in the social, spiritual, ritual, ecological, and weather worlds they share with humans. For this reason, Wouters and Dema argue that as much as life is in lakes—as the latter are integrally life-giving—lakes are also in life, in the sense of consequential agential entanglements between humans and these bodies of water. In Bhutan, lakes are named and personified, manifest as bodies of spiritual instruction and enchantment, and reveal themselves as active protagonists in more-than-human worlds. Anything but inert bodies, *tshomen*-lakes are known to listen, act, create, respond, and disturb— that is, they behave in ways that are considered exclusively human in other places. Anthropogenic changes in lakes, such as their expansion, contraction, or discoloration, are read, experienced, and feared by herders as indicative of a changing relationship between themselves and *tshomen*. Whereas, in the past, *tshomen* were credited with bestowing blessings in the form of wealth, yak progeny, or future visions, today the herders feel that human pollution and sin are driving them away. This is part of a wider problematic in which Himalayan communities blame themselves for changes in the environment and climate that are often primarily caused elsewhere.

If lakes in Bhutan are always more-than-lakes, Himalayan ice is more-than-ice. This is evidenced by Jolynna Sinanan in Chapter 5, on Nepal's Solukhumbu region, which hosts Sagarmatha/Chomolungma/Mount Everest. By innovating cryo-climing and cryo-visual approaches, as an expansion of Sörlin's (2015) cryo-history, which links ice with communities and histories, Sinanan relates Himalayan ice, and changes therein, as "communicative objects" that attach values, narratives, and meanings. Through the lenses of the tourist encounter, labor precarity, digital infrastructure, visual cultures, and the overall global mediatization of Mount Everest, Sinanan merges the local and the global and explores a set of knowing practices of Himalayan ice that move beyond science-only perspectives on the currently changing constellation between ice, verticality, and climate.

Meanwhile, higher up on the Ladakh Plateau, satellite images reveal a number of new lakes that have emerged in the vicinity of the so-called "Line of Actual Control" that forestalls the India–China border dispute. This Line of Actual Control emerges from a deeper, anthropogenic history in the high Himalayas that encompasses British involvement and Tibet's traditional, national geography. Akin to the ice stupas, these lakes are anthropogenic creations: they have been excavated by the Indian Army to store and supply meltwater and groundwater to the thousands of soldiers who patrol the border. India's military build-up in the region hinges on the work of glaciologists and hydrologists tasked with securing a year-round water supply in the arid landscape. Besides digging new lakes and ponds, this effort includes a massive project to revive an ancient lake that seemingly ran dry about 10,000 years ago. The Indian Army's lead geologist commented: "Hydro-geological conditions appear to be conducive for development of [a] paleo channel for groundwater resources … We may have to drill deep but we are confident of finding water for our soldiers" (India Today 2020).

The long-term presence of Indian forces—and, across the Line of Actual Control, their Chinese counterparts—adds further environmental pressure to a landscape that is already vulnerable to anthropogenic climate change. Any military ability to occupy the border on a permanent basis relies on the re-engineering of earth architecture—that is, appropriating and rearranging hydrological and ecological flows. Significantly, the Line of Actual Control is immersed in water as it traverses the landlocked Pangong Tso (lake), which is saline in the east (controlled by India) and fresh in the west (controlled by China). Both countries deploy speedboats and assault vessels to defend their respective territories. In their wake, up and down the lake, these boats disrupt the breeding grounds of migratory birds, including bar-headed geese and Brahmini ducks, as well as the aquatic life of trout, carp, and snails. The border conflict also impacts yaks, sheep, and goats, many of which are cut off from their traditional pastures across the border. Other grazing fields have been turned into buffer zones that are now off-limits to them. Consequently, they are thinner and weaker than they used to be, and their survival relies on the availability of engineered feed supplements brought up from the plains. One herder of Pashima goats lamented, "There has been an increase in mortality among the livestock due to lack of grazing lands; mostly the newborn babies suffer" (Newsclick 2020). Clearly, then, border-making and -patrolling is a more-than-human affair that variously entangles and affects waters, weather and climate, aquatic life, birds, grazing animals, humans, and the built environment. The Line of Actual Control reduces Himalayan diversity and climate in the broadest sense to a bilateral relationship: human geopolitics engender anthropogenic consequences.

This point is forcefully pursued in Chapter 9, where Alexander Davis investigates how an emerging anthropogenic geopolitical clime in which the chief combatants are the Indian and Chinese armies, engineers, and—crucially—the riverine system itself is transforming the Deep Time riverine clime of the Brahmaputra. While classical thinking in geopolitics examines the environmental context as a way of thinking through strategy or as a contested resource, Davis offers a more earth-gravitated and more-than-human understanding of geopolitics that also considers the agency of a river that has refused to be "tamed" by anthropogenic forces.

The mutual constitution between the Brahmaputra's riverine clime and a geopolitical clime began developing when empires sought to border the river and mediate and transform its indigenous flows for state purposes. To some extent, it settled when postcolonial states began solidifying these claims. Today, the Brahmaputra is being engineered, its ecosystems sliced up into bordered, state-based entities, split between India, China, and, to a lesser degree, Bangladesh. State actors are now transforming the river system based on their mutual enmity while damming and transforming the river to fit geopolitical and economic state projects. The riverine clime, for its part, continues to resist these anthropogenic alterations and speaks through floods and other natural disasters. Davis concludes that the relationship between the Brahmaputra riverine clime and anthropogenic states has grown so intimate, transformative, agitated, and critical to contemporary geopolitics between India and China that "the river will wash away the geopolitics, or the geopolitics will destroy the river."

The construction of ice stupas, experiential accounts of lakes changing their behavior, and the pluriverse of Himalayan ice relate how water and climate are experienced, felt, and responded to in material, affective, cognitive, somatic, visual, and spiritual terms in the contemporary, Anthropocenic Himalayas. The more-than-human geopolitics inherent to the India–China border, whether on the Ladakh Plateau or in the Brahmaputra riverine clime, in turn, show the various protagonists, human and not, biotic and not, material and discursive. It further shows the natural and cultural factors in scientific and socio-material practices that are involved in Anthropocenic geopolitics. From an affective earth perspective, these evidence the fragmentation of the Himalayas' integrality. Taken together, they attune to the emplotments, modes of encounter, and anthropogenic pressures, conditions, and transformations of the New Himalayas.

In all of this, the question is not only how anthropogenic activities cause and adapt to Anthropocenic waters and climates. We must also attune to how increasingly Anthropocenic waters are deeply and diversely enmeshed, experienced, and felt in the "everyday" of Himalayan worlds; how these worlds are encountered and framed by climate change; how they are reorienting and reassembling under anthropogenic conditions; and how our meaningful engagement with these worlds may contribute to more complete understandings of Anthropocenic waters and climate in their diversely lived and lively emplacements.

References

Alfthan, Björn. 2017. "The Himalayas—No Longer a 'Blank Spot' for Climate Science." *GRID-Arendal.* https://news.grida.no/the-himalayas-no-longer-a-blank-spot-for-climate-science.

Allison, Elizabeth. 2019. "Deity Citadels: Sacred Sites of Bio-cultural Resistance and Resilience in Bhutan." *Religions* 10(4): 1–17.

Alter, Stephen. 2019. *Wild Himalaya: A Natural History of the Greatest Mountain Range on Earth.* Delhi: Aleph Book Company.

Amrith, Sunil. 2018. *Unruly Waters: How Mountain Rivers and Monsoons Have Shaped South Asia's History.* New York: Basic Books.

Amrith, Sunil and Dan Smyer Yü. 2023. "The Himalaya and Monsoon Asia: Anthropocenic Climes since the 1800s." In *Storying Multipolar Climes of the Himalaya, Andes, and Arctic: Anthropogenic Climate and Shapeshifting Watery Worlds*, edited by Dan Smyer Yü and Jelle J.P. Wouters, 29–51. Abingdon and New York: Routledge.

Banerjee, Milinda and Jelle J.P. Wouters. 2022. *Subaltern Studies: 2.0. Being against the Capitalocene.* Chicago: Chicago University Press/Prickly Paradigm Press.

Barad, Karen. 2007. *Meeting the Universe Halfway: Quantum Physics and the Entanglement of Matter and Meaning.* Durham, NC, and London: Duke University Press.

Bhutia, Kalzang. 2023. "Offerings from the Rivers to the Mountains: Mist and Fog as Connecting Life Force in the Sikkimese Himalayas." In *Storying Multipolar Climes of the Himalaya, Andes, and Arctic: Anthropogenic Climate and Shapeshifting Watery Worlds*, edited by Dan Smyer Yü and Jelle J.P. Wouters, 121–137. Abingdon and New York: Routledge.

Carey, Mark, Philip Garone, Adrian Howkins, Georgina Endfield, Lawrence Culver, Sherry Johnson, Sam White, and James Rodger Fleming, eds. 2014. "Forum: Climate Change and Environmental History." *Environmental History* 19(2): 281–364.

Chakraborty, Ritodhi and Pasang Y. Sherpa. 2021. "From Climate Adaptation to Climate Justice: Critical Reflections on the IPCC and Himalayan Climate Knowledges." *Climatic Change* 167(3–4). https://doi.org/10.1007/s10584-021-03158-1.

Chao, Sophie. 2022. *In the Shadow of the Palms: More-than-Human Becomings in West Papua*. Durham, NC: Duke University Press.

Chao, Sophie and Danielle Celermajer. 2023. "Introduction: Multispecies Justice." *Cultural Politics* 19(1): 1–17.

Chao, Sophie and Dion Enari. 2021. "Decolonising Climate Change: A Call for Beyond-Human Imaginaries and Knowledge Generation." *eTropic: Electronic Journal of Studies in the Tropics* 20(2): 32–54. https://doi.org/10.25120/etropic.20.2.2021.3796.

Chatterjee, Sourav, Muthalagu Ravichandran, Nuncio Murukesh, Roshin P. Raj, and Ola M. Johannessen. 2022. "A Possible Relation between Arctic Sea Ice and Late Season Indian Summer Monsoon Rainfall Extremes." *NPJ Climate Atmospheric Science* 36. https://doi.org/10.1038/s41612-021-00191-w.

Clouse, Carey. 2019. "Climate-Adaptive Design: Building up Ladakh's Ice Stupas." *Landscape Journal: Design, Planning, and Management of the Land* 38(1–2): 25–41.

Crosby, Alfred W. 2004. *Ecological Imperialism: The Biological Expansion of Europe, 900–1900*. 2nd edition. Cambridge: Cambridge University Press.

D'Avignon, Robyn. 2022. *A Ritual Geology: Gold and Subterranean Knowledge in Savanna West Africa*. Durham, NC: Duke University Press.

de Wolff, Kim, Rina C. Faletti, and Ignacio López-Calvo. 2021. *Hydrohumanities: Water Discourse and Environmental Futures*. Oakland: University of California Press.

Delley, Razzeko and Ambika Aiyadurai. 2023. "Eco-Spiritual Water Climes of Dibang Valley, Arunachal Pradesh." In *Storying Multipolar Climes of the Himalaya, Andes, and Arctic: Anthropogenic Climate and Shapeshifting Watery Worlds*, edited by Dan Smyer Yü and Jelle J.P. Wouters, 91–104. Abingdon and New York: Routledge.

Dodds, Klaus and Mark Nuttall. 2015. *The Scramble for the Poles: The Geopolitics of the Arctic and Antarctic*. London: Polity.

Edward, Paul. 2010. *A Vast Machine: Computer Models, Climate Data, and the Politics of Global Warming*. Cambridge, MA: MIT Press.

Eriksen, Thomas H. 2016. *Overheating: An Anthropology of Accelerated Change*. London: Pluto.

Esau, Igor. 2023. "The Arctic Amplification and Its Impact: A Synthesis through Satellite Observations." *Remote Sensing* 15(5): 1345. https://doi.org/10.3390/rs15051354.

Evengård, Birgitta, Joan Nymand Larsen, and Øyvind Paasche, eds. 2015. *The New Arctic*. Cham: Springer. Fleetwood, Lachlan. 2022. *Science on the Roof of the World: Empire and the Remaking of the Himalaya*. Cambridge: Cambridge University Press.

Fleming, James Rodger. 2010. *Fixing the Sky: The Checkered History of Weather and Climate Control*. New York: Columbia University Press.

Gagné, Karine. 2019. *Caring for Glaciers: Land, Animals, and Humanity in the Himalayas*. Seattle: University of Washington Press.

Haberman, David L. 2021. *Understanding Climate Change through Religious Lifeworlds*. Bloomington: Indiana University Press.

Haraway, Donna. 2017. "Symbiogenesis, Sympoiesis, and Art Science Activisms for Staying with the Trouble." In *Arts of Living on a Damaged Planet: Ghosts of the Anthropocene*, edited by Anna Tsing, Heather Swanson, Elaine Gan, and Nils Bubandt, 29–48. Minneapolis and London: University of Minnesota Press.

Huber, Toni. 1999. *The Cult of Pure Crystal Mountain: Popular Pilgrimage and Visionary Landscape in Southeast Tibet*. New York: Oxford University Press.

Hulme, Mike. 2009. *Why We Disagree about Climate Change: Understanding Controversy, Inaction and Opportunity*. Cambridge: Cambridge University Press.

Hulme, Mike. 2013a. *Exploring Climate Change through Science and in Society*. London: Routledge.

Hulme, Mike. 2013b. "How Climate Models Gain and Exercise Authority." In *The Social Life of Climate Change Models: Anticipating Nature*, edited by K. Hastrup and M. Skrydstrup, 30–44. London: Routledge. ICIMOD. 2015. *The Himalayan Climate and Water Atlas: Impact of Climate Change on Water Resources in Five of Asia's Major River Basins*. Kathmandu: ICIMOD.

ICIMOD. 2023. *Water, Ice, Society, and Ecosystems in the Hindu Kush Himalaya: An Outlook*. Kathmandu: ICIMOD.

India Today. 2020. "Ladakh Conflict: Indian Army's Hunt for Water at DBO and Hope to Revive a 10,000-Year-Old Lake." September 16.

Ingold, Tim. 2000. *The Perception of the Environment: Essays in Livelihood, Dwelling and Skill*. London: Routledge.

Ingold, Tim. 2010. "Footprints through the Weather-World: Walking, Breathing, Knowing." *Journal of the Royal Anthropological Institute* 16: S121–S139. https://doi.org/10.1111/j.1467-9655.2010.01613.x.

IPCC. 2007. *Climate Change 2007: Impacts, Adaptation and Vulnerability*. Cambridge: Cambridge University Press.

IPCC. 2013. *The Physical Science Basis*. Cambridge: Cambridge University Press.

Irvine, Richard D.G. 2020. *An Anthropology of Deep Time: Geological Temporality and Social Life*. Cambridge: Cambridge University Press.

Jalais, Annu. 2008. "Unmasking the Cosmopolitan Tiger." *Nature and Culture* 3(1): 25–40.

Jeong, Yong-Cheol, Sang-Wook Yeh, Young-Kwon Lim, Agus Santosa, and Guojian Wang. 2022. "Indian Ocean Warming as Key Driver of Long-Term Positive Trend of Arctic Oscillation." *NPJ Climate and Atmospheric Science* 56. https://doi.org/10.1038/s41612-022-00279-x.

Kimmerer, Robin W. 2014. *Braiding Sweetgrass: Indigenous Wisdom, Scientific Knowledge, and the Teachings of Plants*. Minneapolis: Milkweed Press.

Klaver, Irene J. 2022. "Radical Water." In *Hydrohumanities: Water Discourse and Environmental Futures*, edited by Kim de Wolff, Rina C. Faletti, and Ignacio López-Calvo, 64–90. Oakland: University of California Press.

Knox, Hannah. 2020. *Thinking Like a Climate: Governing a City in Times of Environmental Change*. Durham, NC, and London: Duke University Press.

Kohn, Eduardo. 2013. *How Forests Think: Toward an Anthropology beyond the Human*. Berkeley: University of California Press.

Kumar, Vikash, Manish Tiwari, Dmitry V. Divine, Matthias Moros, and Arto Miettinen. 2023. "Arctic Climate-Indian Monsoon Teleconnection during the Last Millennium Revealed through Geochemical Proxies from an Arctic Fjord." *Global and Planetary Change* 222. https://doi.org/10.1016/j.gloplacha.2023.104075.

Kumari, Nikhul, Ankur Srivastava, and Umesh Chandra Dumka. 2021. "A Long-Term Spatiotemporal Analysis of Vegetation Greenness over the Himalayan Region Using Google Earth Engine." *Climate* 9(7): 109. https://doi.org/10.3390/cli9070109.

Langenbrunner, Baird. 2020. "Glacial Lakes: Hazards in the Himalayas." *Nature Climate Change* 10(385). https://doi.org/10.1038/s41558-020-0778-0.

Ma, Jieru, Tinghan Zhang, Xiaodan Guan, Xiaoming Hu, Anmin Duan, and Jingchen Liu. 2019. "The Dominant Role of Snow/Ice Albedo Feedback Strengthened by Black Carbon in the Enhanced Warming over the Himalayas." *Journal of Climate* 32(18): 5883–5899. https://doi.org/10.1175/jcli-d-18-0720.1.

Manish, Kumar and Maharaj K. Pandit. 2018. "Geophysical Upheavals and Evolutionary Diversification of Plant Species in the Himalaya." *PeerJ* 6: e5919. https://doi.org/10.7717/peerj.5919.

Mathur, Nayanika. 2015. "'It's a Conspiracy Theory *and* Climate Change': Of Beastly Encounters and Cervine Disappearance in Himalayan India." *Hau: Journal of Ethnographic Theory* 5(1): 87–111.

McNeill, John R. and Matthew Engelke. 2016. *The Great Acceleration: An Environmental History of the Anthropocene since 1945*. Cambridge, MA: Harvard University Press.

Morton, Timothy. 2013. *Hyperobjects: Philosophy and Ecology after the End of the World*. Minneapolis and London: University of Minnesota Press.

Newsclick. 2020. "Nomadic Pastoralists in Ladakh Face 'Exodus' as Indo-China Tension Spikes along LAC." October 4. www.newsclick.in/nomadic-pastoralists-ladakh-face-exodus-Indo-China-tension-spikes-LAC.

Oberborbeck Andersen, Astrid. 2023. "Pluriversal Tundra: Storying More-than-Human Ecologies across Deep, Accelerated, and Troubled Times." In *Storying Multipolar Climes of the Himalaya, Andes, and Arctic: Anthropogenic Climate and Shapeshifting Watery Worlds*, edited by Dan Smyer Yü and Jelle J.P. Wouters, 69–87. Abingdon and New York: Routledge.

Paerregaard, Karsten. 2023. "Climing the Andes: Vertical Complementarity, Transhuman Reciprocity, and Climate Change in the Peruvian Highlands." In *Storying Multipolar Climes of the Himalaya, Andes, and Arctic: Anthropogenic Climate and Shapeshifting Watery Worlds*, edited by Dan Smyer Yü and Jelle J.P. Wouters, 53–68. Abingdon and New York: Routledge.

Pandit, Maharaj K. 2017. *Life in the Himalaya: An Ecosystem at Risk*. Cambridge, MA, and London: Harvard University Press.

Pant, G.B., Pradeep P. Kumar, Jayashree V. Revadekar, and Narendra Singh, eds. 2018. *Climate Change in the Himalayas*. Cham: Springer International Publishing.

Parvaiz, Athar. 2018. "Ice Stupas Spark Conflict between Farmers in Ladakh." *The Third Pole*, July 5. www.thethirdpole.net/en/climate/ice-stupa-ladakh/.

Povinelli, Elizabeth A. 2016. *Geontologies: A Requiem to Late Liberalism*. Durham, NC, and London: Duke University Press.

Previdi, Michael, Karen L. Smith, and Lorenzo M. Polvani. 2021. "Arctic Amplification of Climate Change: A Review of Underlying Mechanisms." *Environmental Research Letters* 16(9): 1–26. doi:10.1088/1748-9326/ac1c29.

Saikia, Arupjoyti. 2019. *The Unquiet River: A Bibliography of the Brahmaputra*. Oxford: Oxford University Press.

Scientific American. 2010. "How Fast Are Himalayan Glaciers Melting?" January 21. www.scientificamerican.com/podcast/episode/how-fast-are-himalayan-glaciers-mel-10-01-21/.

Searle, Mike. 2013. *Colluding Continents: A Geological Exploration of the Himalaya, Karakoram, and Tibet*. Oxford: Oxford University Press.

Serreze, Mark. 2018. *Brave New Arctic: The Untold Story of the Melting North*. Princeton: Princeton University Press.

Shekhar, M.S., H. Chand, S. Kumar, K. Srinivasan, and A. Ganju. 2010. "Climate-Change Studies in the Western Himalaya." *Annals of Glaciology* 51(54): 105–112.

Shepherd, Theodore G. et al. 2018. "Storylines: An Alternative Approach to Representing Uncertainty in Physical Aspects of Climate Change." *Climate Change* 151: 555–571.

Shepherd, Theodore G. and Chi Huyen Truong. 2023. "Storylining Climes." In *Storying Multipolar Climes of the Himalaya, Andes, and Arctic: Anthropogenic Climate and Shapeshifting Watery Worlds*, edited by Dan Smyer Yü and Jelle J.P. Wouters, 157–183. Abingdon and New York: Routledge.

Shroder, John F., ed. 1993. *Himalaya to the Sea: Geology, Geomorphology and the Quaternary.* London and New York: Routledge.

Singh, Vandana. 2023. "Not Just the Science: A Transdisciplinary Pedagogy for Cryospheric Climes." ." In *Storying Multipolar Climes of the Himalaya, Andes, and Arctic: Anthropogenic Climate and Shapeshifting Watery Worlds*, edited by Dan Smyer Yü and Jelle J.P. Wouters, 184–200. Abingdon and New York: Routledge.

Smyer Yü, Dan. 2014. "Sentience of the Earth: Eco-Buddhist Mandalizing of Dwelling Place in Amdo, Tibet." *Journal for the Study of Religion, Nature, and Culture* 8 (4): 483–501. doi:10.1558/jsrnc.v8i4.19481.

Smyer Yü, Dan. 2020. "The Critical Zone as a Planetary Animist Sphere: Ethographing an Affective Consciousness of the Earth." *Journal for the Study of Religion, Nature, and Culture* 14(2): 271–290. http://doi.org/10.1558/jsrnc.39680.

Smyer Yü, Dan. 2021. "Situating Environmental Humanities in the New Himalayas: An Introduction." In *Environmental Humanities in the New Himalayas: Symbiotic Indigeneity, Commoning, Sustainability*, edited by Dan Smyer Yü and Erik de Maaker, 1–21. Abingdon: Routledge.

Smyer Yü, Dan. 2023. "Multipolar Clime Studies of the Anthropocenic Himalaya, Andes and Arctic: An Introduction." In *Storying Multipolar Climes of the Himalaya, Andes, and Arctic: Anthropogenic Climate and Shapeshifting Watery Worlds*, edited by Dan Smyer Yü and Jelle J.P. Wouters, 1–26. Abingdon and New York: Routledge.

Smyer Yü, Dan and Jelle J.P. Wouters, eds. 2023. *Storying Multipolar Climes of the Himalaya, Andes and Arctic: Anthropogenic Climate and Shapeshifting Watery Worlds.* Abingdon and New York: Routledge.

Snively, Gloria and John Corsiglia. 2001. "Discovering Indigenous Science: Implications for Science Education." *Science Education* 85(1): 6–34.

Sörlin, Sverker. 2015. "Cryo-History: Narratives of Ice and the Emerging Arctic Humanities." In *The New Arctic*, edited by Birgitta Evengård, Joan Nymand Larsen, and Øyvind Paasche, 327–339. New York: Springer.

Sturm, Matthew, Donald K. Perovich, and Mike C. Serreze. 2003. "Meltdown in the North." *Scientific American* 298(4): 60–67. doi:10.1038/scientificamerican1003-60.

Sundaram, Suchithra and David M. Holland. 2022. "A Physical Mechanism for the Indian Summer Monsoon—Arctic Sea-Ice Teleconnection." *Atmosphere* 13(4): 1–15. https://doi.org/10.3390/atmos13040566.

Tsosie, Krystal S. and Katrina G. Claw. 2020. "Indigenizing Science and Reasserting Indigeneity in Research." *Human Biology* 91(3): 137–140.

van Dooren, Tom and Deborah Bird Rose. 2016. "Lively Ethography: Storying Animist Worlds." *Environmental Humanities* 8(1): 77–94. https://doi.org/10.1215/22011919-3527731.

van Schendel, Willem. 2009. *A History of Bangladesh.* Cambridge: Cambridge University Press.

Walsh, James and Christopher Rackauckas. 2015. "On the Budyko-Sellers Energy Balance Climate Model with Ice Line Coupling." *Discrete and Continuous Dynamical Systems—Series B* 20(7): 1–10. doi:10.3934/dcdsb.2015.20.xx.

Walther, Michael, Avirmed Dashtseren, Ulrich Kamp, Khurelbaatar Temujin, Franz Meixner, Caleb G. Pan, and Yadamsuren Gansukh. 2017. "Glaciers, Permafrost and Lake Levels at the Tsengel Khairkhan Massif, Mongolian Altai, during the Late Pleistocene and Holocene." *Geosciences* 7(3). https://doi.org/10.3390/geosciences7030073.

Wang, Liping, Haijun Yang, Qin Wen, Yimin Liu, and Guoxiong Wu. 2023. "The Tibetan Plateau's Far-Reaching Impacts on Arctic and Antarctic Climate: Seasonality and Pathways." *Journal of Climate* 36(5): 1399–1414. https://doi.org/10.1175/jcli-d-22-0175.1.

Wester, Philippus, Arabinda Mishra, Aditi Mukhergi, and Arun Bhakta Shrestha. 2019. *The Hindu Kush Himalaya Assessment: Mountains, Climate Change, Sustainability and People.* Cham: Springer International Publishing.

Wouters, Jelle J.P. 2023a. "Conclusion: Multilateral Clime Studies." In *Storying Multipolar Climes of the Himalaya, Andes, and Arctic: Anthropogenic Climate and Shapeshifting Watery Worlds*, edited by Dan Smyer Yü and Jelle J.P. Wouters, 253–272. Abingdon and New York: Routledge.

Wouters, Jelle J.P. 2023b. "Where is the 'Geo'-political? More-than-Human Politics, Polities, and Poetics in the Bhutan Highlands." In *Capital and Ecology Developmentalism, Subjectivity and the Alternative Life-Worlds*, edited by Rakhee Bhattacharya and G. Amarjit Sharma, 181–202. Abingdon: Routledge.

Wouters, Jelle J.P. and Michael T. Heneise. 2023. "Highland Asia as a World-Region: An Introduction." In *The Routledge Handbook of Highland Asia*, edited by Jelle J.P. Wouters and Michael T. Heneise, 1–40. Abingdon and New York: Routledge.

Xu, Jianchu, R. Edward Grumbine, Arun Shrestha, Mats Eriksson, Xuefei Yang, Yun Wang, and Andreas Wilkes. 2009. "The Melting Himalayas: Cascading Effects of Climate Change on Water, Biodiversity, and Livelihoods." *Conservation Biology* 23 (3): 520–530. https://doi.org/10.1111/j.1523-1739.2009.01237.x.

Yusoff, Kathryn. 2013. "Geological Life: Prehistory, Climate, Futures in the Anthropocene." *Environment and Planning D: Society and Space* 31: 779–795.

Zee, Jerry C. 2022. *Continent in Dust: Experiments in Chinese Weather System.* Berkeley: University of California Press.

2 Paddy Clime

Ecological Indigeneity in the Naga Uplands

Roderick Wijunamai

One late morning in December 2020, I was sunbathing on the front porch of my house. My family was about to have our lunch, and my mother had gone into the kitchen to get our meal ready. A paternal cousin from my ancestral village suddenly showed up with a half-filled polypropylene cement sack on her back. "It is *meusan* [freshly harvested rice]," she said. "I just wanted your mother to taste our harvest." The cousin had grown up in my house, and my mom had looked after her until she married another villager.

When my mother came out with our meal, she was pleased to see the gesture. Nothing excites her more than *meusan* as a gift, especially during Christmastime. "I can deal with bad curry; I can even go without any curry and just eat chutney with rice if the rice is good," she would often say. Cooking "good rice" was important for her. Nowadays, our storeroom, where we stock our months-long supplies, is more often loaded with "superfine rice"—a foreign rice variety, primarily sold and supplied by the Food Corporation of India, a statutory organization under the Government of India. "Local *baan*" (*baan* meaning rice, a short form of *chaban*), as we call it, is not readily available, and even if it can be found, it is priced very high in the market. The replacement of "indigenous rice" with "superfine rice" marks a transformation in Naga agricultural and food history. It is a trend that worries my mother and many other Nagas.

I belong to the Liangmai Naga community that inhabits the Southern Naga Hills, which today are divided by the political border of Manipur and Nagaland. Growing up, I learned to distinguish between local rice and non-local varieties (of which there are many). Local rice has a special place in our community. It confers respect, recognition, and blessing. Food writ large is something we consume not just for our physical health, but also for our cultural and spiritual well-being. As such, food is sacred, and treated with reverence, from sowing to harvesting and gathering, and from preparation to consumption.

Naga elders often instruct their children not to eat "improper food" (referring to industrial, packaged food and other cuisines from outside the region), and encourage them to eat rice instead. Parents are happy when their children comply. "Rice is the progeny of God; or, perhaps, after God is rice, and after rice is wealth," Gaidon, a 75-year-old Rongmei man, told my friend Gairan during

DOI: 10.4324/9781003484394-2

an interview.[1] Time and again, as children, we would hear elders making similar comments about rice, by which they meant our indigenous rice. Here, "indigenous" refers to rice that is grown on Naga soil, and sown, transplanted, harvested, and cooked by Naga hands and techniques. Anthropogenic transformations—of both human–land relations, including the move away from agriculture, and weather and climate patterns—are now undermining the position of Naga indigenous rice.

Introduction: A Paddy Clime

This chapter invites its readers to the Naga uplands to reflect on the multispecies and multidimensional relations of paddy in the making and unmaking of more-than-human worlds. In geological terms, the Naga hills, valleys, and mountains are part of a longer ophiolite belt, sometimes referred to as the "Naga–Andaman suture," which runs from present-day Nagaland and Manipur in India's Northeast, through western Myanmar, all the way to the Andaman and Nicobar Islands in the Indian Ocean (Baxter et al. 2011). Geologists argue that the Naga ophiolite—the emplaced oceanic crust in the Naga lands—alongside other ophiolites in the Alpine–Himalayan Orogenic Belt, is one of the "most comprehensive ophiolite depositories in earth's history" (Furnes et al. 2020, 1). In other words, the Naga geography is part of a belt that houses one of the most comprehensive archives of our geological history. This Deep Time geological history antecedently provided the material setting for the contemporary paddy clime of the Naga uplands.

I conceive of a "clime" as the affective emplacement, embodiment, and entwinement of earth, human, vegetal, animal, and elemental actants, relations, and imaginaries. Grounded in ethnography as well as in my own embodied experience as a Naga, this chapter brings the material-semiotic "plant turn" in the environmental humanities, and allied post-humanist currents, into conversation with emergent clime studies (Smyer Yü and Wouters 2023) through the storying and enlivening of a "paddy clime."[2] Paddy and humans partake in an ongoing multispecies story that has transformed both. Paddy is among the most successful plants of our multispecies history. Its ability to nourish and grow human bodies allowed it to garner dominant status in the plant world, occupying and governing surface worlds, ecosystems, and environmental flows. Humans have variously adapted their livelihoods and lifeworlds to the whims and wills of this plant. My coinage of "paddy clime," and as enlivened in the Naga uplands, affectively recognizes the centrality of paddy in Naga histories up to the unfolding present as well as its integrality to local ecosystems, soil, environmental flows (particularly of water), multispecies worlds, and cultures, and to Nagas' affective and spiritual relations with the fecundity of paddy fields. Among the Naga, paddy occupies a central place in the physical environment, along with its association with social, cultural, political, and psychological worlds. Climate/clime change, in turn, is experienced in terms of changing paddy–human relations.

While climate science knowledge privileges secular, technical, and human-focused understandings of climate (change), clime studies calls for more capacious accounts that affirm climate change and its various environmental crises as always a multispecies affair (Wouters 2023). Among Nagas, anthropogenic transformations, including both climate change and the workings of the capitalist market, are changing the fecundity of the soil, precipitation patterns, the production of local crops (especially paddy), and consequently the food, bodies, and experienced well-being of Naga villagers. Put differently, climate change forces many Nagas to eat rice that is not grown on their own soil, which I term "non-indigenous rice." This has repercussions that are at once physical, emotional, cultural, and spiritual. In this context, I explore the affective, relational, and terrestrial lives of paddy and the myriad ways it participates in the making and unmaking of more-than-human climes and contemporary anthropogenic changes therein. By coining the affective notion of paddy clime, I foreground the role of paddy in the social behavior, ritual lives, bodies, health, and psyche of Nagas.

At a broad level, this chapter explores the agential relationships between paddy and humans in a particular place—the Naga uplands. It presents paddy as an active protagonist of Naga history by tracing its origin and arrival in Naga lands, its imperial expansion by acquiring dominance in the biosphere, its reorienting of indigenous water flows through its semi-aquatic cultivation (which was promoted by colonial and postcolonial governments), and its various bearings on Naga political, economic, cultural, and spiritual domains. In clime terms, this chapter affirms paddy as a clime-maker due to its impact on the terrestrial and social landscapes. Clime change, in turn, is experienced in alterations in the co-constitutive relations—of soil, rain, waters, farmers, rituals, and culture—that allowed paddy to dominate the local landscape for centuries. In particular, I foreground the local experience of climate change by detailing the culturally–spiritually affirmed indigenous knowledge of rain typologies that is key in paddy clime, and which today is disturbed. In short, this chapter proposes understanding climate change's "slow violence" through the lens of the paddy in the Naga uplands.[3] I ask, what does climate change look like from a paddy point of view?

In their attempt to theorize human–fish relations among the Indigenous peoples of Canada, the Métis anthropologist Zoe Todd (2014) introduced the idea of "fish pluralities." According to Todd (2014, 217), these are the "multiple ways of knowing and defining fish," as invoked by the Inuvialuit of Paulatuuq, "to negotiate the complex and dynamic pressures faced by humans, animals, and the environment in contemporary Arctic Canada." To that end, in the case of the Naga uplands, I consider the possibility of "paddy pluralities," in terms of the complex and many ways Nagas think of, relate to, and are subjected by paddy. In the pages that follow, I conflate this conception with the idea of "clime" as articulated by Fleming (2010) and Carey and Garone (2014), and revived and expanded by Smyer Yü and Wouters (2023). I describe "paddy clime" to show the dynamic relationship between humans and paddy in an emergent form of life. Paddy clime variously affirms paddy's affective role in structuring and transforming ecological, hydrological, economic, political, cultural, and spiritual realm of the Naga lifeworlds. This is also along the lines of Candis Callison's contention that climate change "is not a straightforward problem nor is it a standalone fact" (Callison 2014, 12). Clime studies, then, is a response to such complexities; it presents itself as a pertinent forum in its focus on situated, terrestrial, experiential, and multispecies entanglements of place and climate history, pattern, and change.

Specifically, this chapter draws on insights and stories I gathered from interlocutors and kin across twelve Naga tribes who self-identify under the umbrella term Tenyimi (or Tenyimia), a larger community of which I am also a member.[4] In the following pages, I invoke and engage Tenyidie idiom and axioms relating to rice, paddy, and rain. (The members of the Tenyimi community trace their origins to a person called Tenyiu. Tenyi-mi or Tenyi-mia simply means "Tenyi-people," while Tenyi-die translates as "Tenyi-language.") To be sure, the cultures among the Tenyimi are far from homogeneous, and there are marginal differences in our lifeways. Even so, I argue that the larger construct and articulation of interspecies relations can be understood in some unison, and can contribute to our understanding of climes and ecological crises from the vantage point of the Naga rice paddy, thus allowing a more-than-human and multidimensional perspective on climate change.

Putting Rice in Naga History

Archaeologists gather around the conviction that rice paddy (*Oryza sativa*) was first cultivated in the Yangtze River Basin in China somewhere between 8000 and 13,500 years ago (Callaway 2014; Fornasiero, Wing, and Ronald 2022). The material ground for a worldwide transition in land use and human diet was thus set. Our planet became a rice-growing earth, soil adjusted to paddy, paddy adjusted to soil—later, soil was fertilized and changed to subdue it to the wishes of paddy—and the human cultural, spiritual, and political world formed around paddy. Across Asia, Africa, and the Americas, paddy climes expanded with human movement and migration, integrating themselves within the local ecologies wherever they went. Colonial expansion, war, migration, and trade spread paddy around the world, first across Asia, then further afield, including to the Americas as part of the "Columbian Exchange" (Carney 2001). Today, more than half of the world depends on rice as a staple crop.

Emergent debates in archaeo-botany and linguistics in Northeast India are about whether paddy is indigenous or a later settler in the region (Hazarika 2023). What is beyond doubt is that, apart from humans, paddy has turned into the dominant plant species in the region, shaping its histories, polities, and politics, closely followed by the tea plant, which has received much more local and international scholarly attention (Sharma 2011; Besky 2014; Dey 2018; Karlsson 2022). Archaeologists and biologists are increasingly confident about the origins of paddy cultivation in the Yangtze River Basin; however, it is quite possible that it was independently domesticated and cultivated in different geographies at different times, including in the Naga uplands.

According to the linguist George van Driem (2017), the Brahmaputra Valley and its adjacent hills were one of the first places where rice paddy was cultivated and then, many years later, domesticated (see also Hazarika 2017). Arguably, this cultivation first took place in the hills, followed by the valley. This argument is substantiated by the historian Arupjyoti Saikia (2019), who notes that livelihood sustenance in the hills was comparatively easier and necessitated less labor, whereas valley-dwellers were subject to constant flooding and were more vulnerable to enemy attack (cf. Ludden 2005). Others (e.g., Okoshi et al. 2018; Choudhury, Latif Khan, and Dayanandan 2013) have suggested that the Eastern Himalayas, along with Yunnan in China and other Southeast Asian highlands, are the "gene center" of *Oryza sativa*, as they are home to a wide variety of indigenous rice. In Nagaland state alone, excluding large tracts of Naga uplands outside the state, the State Agriculture Research Station has documented 867 landraces of rice (Government of Nagaland 2017). The presence of such high variety is arguably due to the region's topographic features, with varied geo-ecological gradients and niches, which means it is host to differential precipitation patterns, temperatures, and biodiversity.

Ethnolinguists, archaeologists, and historians propose that the peopling of the hills of Northeast India was a protracted process in which humans arrived from different directions at different times. For many centuries, movement was key to human presence in the region, and slash-and-burn (or swidden) cultivation testified to this movement. This is also substantiated by origin stories, many of which talk about early and frequent migration, with some communities tracing their origins as far as East Asia (Huber and Blackburn 2012). Arguably, it was paddy that allowed people to settle and form communities. Pachuau and van Schendel (2022) argue that rice paddy was the "most intrusive" crop in the region. It might as well have been a symbiotic intrusion, with humans and paddy collaborating to take over the hills.

The arrival of rice not only made Naga life more sedentary but also drastically changed the ecology through the privileging of paddy. In the Naga uplands, the cultivation of rice paddy can be broadly classified into wet-rice cultivation and dry-rice cultivation. Most Naga tribes cultivated dry paddy through *jhum* (swidden agriculture) until relatively recently. When the British first arrived in the Naga Hills, they recorded only the Angami—then including the Chakhesang as "Eastern Angamis"—and a few adjacent Zeliangrong villages,[5] on the southern side of the hills, as practicing wet-rice cultivation (Hutton 1921). After seeing the terrace cultivation of the Angamis in 1873, the English topographer Henry Haversham Godwin-Austen noted, "I have never, even in the better-cultivated parts of the Himalayas, seen terrace cultivation carried out to such perfection, and it gives a peculiarly civilized appearance to the country" (cited in Elwin 1969, 587–588). Likewise, the following year, the political agent in Manipur, R. Brown, reported: "The labour incurred in making these terraces must be very great, and the skill manifested in irrigating them would do great credit to a trained engineer" (cited in Elwin 1969, 588–589). Indeed, wet-rice cultivation entails the anthropogenic shaping and sculpting of the landscape, including the containment of environmental flows of water so that it remains in the paddy fields.

The political scientist James Scott (2009) argues that the Southeast Asian highland peoples—ironically including those in India's Northeastern highlands—consciously avoided state control by practicing slash-and-burn cultivation and growing crops that "are of staggered maturity, fast growing, and easily hidden, if they require little care, are of little value per unit weight and volume, and grow below ground, they acquire greater escape value" (Scott 2009, 199). In contrast to wet-rice cultivation, which can hardly be hidden, slash-and-burn agriculture had "escape value" because fields were always on the move. Nagas, for the most part, do not describe either their choice of crop or slash-and-burn agriculture in political terms as helping them avoid the state historically. Nonetheless, their cultivation system did indeed frustrate colonial administrators, as it made the work of governing the region more difficult. Hence, colonial officers designed policies that sought to change it (Das 2018). Rice paddy, in this sense, not only subjugated other crops but also made the Nagas more governable. Paddy territorialized the Naga uplands and became a prime medium through which the state could exert control.

Above all, the cunning of paddy manifests in the way Nagas organize their lives and lifeworld. From our notion of time to our understanding of the seasons and weather, our idea of ethics and justice, and even our conception of life after death, rice made its way to the center of the Naga world. Spiritual offices were enacted around its cultivation. Ownership patterns of land emerged. Social organization of clans became attuned to it. Festivals celebrated different parts of the annual paddy cycle. Notions of wealth became defined in relation to paddy (see discussion of the "feast of merit," below). Political subjugation, whether to the British or to neighbors, came to be measured in rice tax, and inter-tribal conflicts that continue to this day may be traced to the demand of rice. Clearly, then, rice etched a place for itself at the center of Naga social, political, cultural, economic, and spiritual life. While paddy once thrived in a multispecies setting, its settled, wet-rice cultivation means it is now isolated. In *jhum*, agricultural fields were environments where many species—including humans and soils—symbiotically nourished, sheltered, and provided for each other. Land was left fallow for a few years and it replenished. Ornithologists—using bird species as an indicator of biodiversity—confirm that *jhum* supports more lifeforms than artificially forested ecosystems (Mandal and Shankar Raman 2016).

Yet, over time, paddy became increasingly semi-aquatic and started to strike up new relations with water. It also found some new partners, especially snails, which, in turn, etched themselves into the heart of the Tenyimi diet. Because of this transition, the environmental flow of water became anthropogenically altered. Elaborate projects to retain water in the hills and stop it from flowing down to the plains included the building of terraces—back-breaking work that consumed tremendous amounts of Naga labor and demanded ongoing maintenance and repair. Irrigation channels, in turn, were crucial to survival, so ownership of them and access to them became a central part of Naga customs.

To be sure, rice paddy was not the predominant crop throughout the Naga uplands prior to the mid-twentieth century. Since then, though, both colonial and postcolonial governments have actively propagated rice as a "civilized" crop (Das 2020). The authorities disapprove of dry, shifting cultivation because the yields are comparatively small and the fields can be hidden in the jungle, far from government control, both of which result in less revenue. The government-facilitated turn, which continues to this day, not only sparked political changes among the Naga, making their produce more visible, but also changed the nature of the paddy plant itself.

Today, many Naga tribes, especially those who did not cultivate rice until recently, associate crops like millet with "the days of poverty." The following is an extract from a conversation I had with my Konyak Naga host during a trip to the easternmost part of the Naga uplands, in the Indo-Myanmar borderlands:

AUTHOR: I want to try *shi* [millet].
HOST: [Laughs.] Why?
AUTHOR: I just want to try it. Don't you have it at home? I saw your neighbor having it today.
HOST: Of course we do. We eat it sometimes.
AUTHOR: Shall we make it for dinner?
HOST: [Laughs again.] No. We cannot cook millet for guests. You are our guest.

Certainly, my host was half joking, and I eventually managed to convince my host to cook millet that evening. But my host made it clear that it was inauspicious to do so for guests. It may not always have been the case, but rice had come to symbolize prosperity and blessings.

Affective Human–Paddy Relations

Among many Naga communities, paddy arranged the social order and divisions within some clans, *khel* (village wards), and regions explicitly associated with rice varieties. Terrace cultivation channels perennial water from field to field, and because community lands are divided across clans and *khel*, different fields produce different harvests, which in turn become identity markers in some Naga villages. Kethoseno, an Angami woman in her eighties, explained:

> The water is cold on the upper side of the terrace cultivation, and as it flows down, the water gets warmer. Because of this, people get different paddy plants, as well as different variants of rice, depending on where their field is located.

In an indigenous paddy clime, then, villagers are able to differentiate paddy by its gradients and niches. Depending on where each clan or *khel* is allocated land, different varieties of rice are produced, and attachments ensue.

That Nagas and rice paddy are thickly intertwined in an intersubjective relationship is best understood through the intimate rituals surrounding the paddy harvest. After the arrival and spread of Christianity, some of these rituals fell into decline, were forgotten, or adapted themselves to Christian worldviews. They are, however, remembered in oral histories and transmitted by contemporary Naga novelists. Easterine Kire, a well-known storyteller, has arguably done the most work in transmitting Naga spiritual and cultural worlds through her novels, poems, and short stories. In her novella *Don't Run, My Love*, Kire (2017a) tells the story of a village harvest that is delayed by a week because the *liedepfü* (a ritual harvest initiator) is ill. I asked Kire to elaborate on the person who performs this role. She explained:

> In our culture, the process of growing food and harvesting it, and the long period of tending it in between, is approached in a solemn manner. The role of *liedepfü* can only be done by a woman. She formally opens the harvesting season. She would take a scythe and walk into the ripe fields and harvest a few sheaves of grain into her basket. Once that is done, the rest of the community can begin harvesting … To commence harvesting without the ritual performed by the *liedepfü* is to invite disaster in different forms. Perhaps the grain won't be as long-lasting. Perhaps rodents might consume the grain, or a sudden invasion of the village could happen, resulting in the destruction of lives and property. Inclement weather could also destroy stocks of grain when the right rituals have not preceded harvesting.

Analogous to this explanation, more than a century ago, the colonial anthropologist John Henry Hutton (1921, 189) mentioned the *liedepfü*'s male counterpart, the *tsakro*, who initiated sowing among the Angami. It was taboo for anyone in the village to begin either sowing or harvesting before the ritual initiator had formally inaugurated the field. There are various names for these positions across Tenyimi communities, but the rituals and observances they perform are very similar.

It is considered crucial that the holders of these ritual offices are "pure," "healthy," and "clean," as defined in cultural terms. There is a Rongmei folk song about a certain Mathuthuanlu—an allegedly "impure" (menstruating) woman—who attempted to perform the ritual and disappeared on her way back home, never to be seen again. A village elder told me that it is taboo for any person—be they male or female—who is "impure" or has an "unclean" history to perform the ritual. It is also believed that this precondition is necessary to learn and understand how to read the weather, the moon, and other signs.

"It may seem like the women's role in the affective relations and agricultural rituals are inconsequential," Senganglu, a Rongmei academic told me, but "Naga women are the actual 'bread-earners' of their families. Men are traditionally warriors; even in old pictures, they will be seen with their *dao* [Naga knives]. Women are the ones most engaged in food gathering or gardening."

Senganglu and other Tenyimi elders repeatedly referred to an important ritual that is performed by the mother of the house during harvest festival. Again, Kire provided further details:

> On the day after the harvest has been brought home, she [the mother] will perform the ritual to make her grain long-lasting. She stays in the house all day and eats lentils, millet, maize, pumpkin, and other vegetables found around the time of paddy harvesting. The belief is that these grains and vegetables will be made long-lasting by the mother's observance of this ritual.

What significantly defines multispecies life in a paddy clime is that ritual and sacralization are arranged around the annual cycle of sowing, transplanting, harvest, and eating paddy. The performance of these rituals, however, includes other participating species as well. *Baan-lao-laomei*—one of the primary Rongmei rituals revolving around paddy—is conducted at the village gate by every family before going to the fields to start the first phase of planting rice. Two distinct lines of small holes are made, and seeds are pushed into them. One line is an offering to the creator deity for blessing, and another for insects and pests, as their share, so that they will not attack or hinder the rice plants.

Kenyü is a strong disciplinary force within the community as it is believed that harm comes to anyone who does not follow it. (*Kenyü* is an Angami term; other Tenyidie dialects have different words for the same concept.) Pertaining to seed grains, at the time of harvesting, *kenyü* tells us that we must not pick up any grains that have fallen in the field, as they must be left for birds and insects. Similarly, whenever food falls out of the *unuva* (packed lunch), it is taboo to pick it up. "We are not supposed to take back what the wind has taken away," Meno, an Angami friend explained to me. However, this does not mean Nagas waste seeds or food. Rather, Meno repeated what her grandmother once told her:

> It is taboo to destroy seed-grain … Grain is seen as a sign of being blessed by the creator deity. To possess full granaries means you possess the blessing of the creator deity. Throwing food, or playing with food, is a very strict taboo. It is considered sacrilegious. It angers the creator deity because it reflects an attitude that does not appreciate his gifts. Violation of the taboo always brings punishment. The village that wastes grain with full knowledge will be punished with famine.

Clearly, then, rice is sacrosanct for Nagas and, therefore, held ceremoniously in our collective conscience and lifeways. The divine weight of rice in the

Naga lifeworld is also expressed through shamanic rituals in which it plays the cardinal role as a divine agent. Angami shamans use four or seven grains of rice in their divination rituals, during which spirits communicate to the shamans through the grains (Joshi 2004).

Paddy also became central to Nagas' "prestige economy" (Lehman 1989). A good deal has been written about the Naga "feast of merit" (e.g., Stonor 1950; Wouters 2015; Yekha-ü and Marak 2021; Mayirnao and Khayi 2023), although most of these accounts focus on its political or social dimensions, rather than considering it from a multispecies perspective. Christoph von Fürer-Haimendorf coined the phrase in his monograph *The Naked Nagas* (von Fürer-Haimendorf 1935), while a British colonial officer, James Philip Mills (1935, 134), provided the first detailed description of what it entailed:

> [T]he essence of this system is that every male Naga, if he is to acquire merit and status in this world and the next, must give a series of feasts, every detail of which is strictly prescribed … An immense amount of rice and rice beer is consumed, and more rice than money goes to the buying of the animals … More important still are the privileges they win for the giver.

These privileges—or signs of prestige—include special embroidery, carved posts, and monoliths erected in the feast-giver's name.

There are three parallel worlds in Zeliangrong Naga cosmology: the land of gods (*tingkao-ram*), the land of death (*tarui-ram*), and the corporeal world (*nban-ram*). The latter is inhabited by humans (in their current bodily form), spirits, animals, birds, and other species. Most humans have to pass through the land of death in order to reach the land of gods. However, prosperity provides a direct path to the land of gods. In the grammar of Zeliangrong society, the wealthiest person is the one with the largest granary. So, the best way to achieve prosperity is to reap recurrent bumper harvests. When any "divinely blessed" woman or a man (who is not a widow or a widower) reaps a bountiful harvest, they are expected to host an almost month-long ritual (*banrü*) and feasting. The bountiful harvest, I was told, should be "something people in the village have never seen," and the blessing is to be shared before moving to the land of gods. This is the most prestigious Naga ritual, and the most sacred feast. In essence, ultimate success depends on a fruitful interspecies relationship between the Naga and their rice paddy.

This spills over into dreams, too (see Heneise 2018). I grew up learning that it is a good omen to dream of a ripened field or baskets of grain, and that "blessings of abundance will follow the dreamer." These are all ecological expressions of the sacred relations Nagas have with paddy. They show that the Naga ecology has been attuned to the wishes and needs of paddy, with human habitation and cultural, spiritual, and multispecies relations revolving strongly around the plant. What is more, it is through these expressions that upland Nagas sustain their symbiotic indigeneity not just with paddy but also with the indigenous other-than-humans.

More-than-Human Indigeneity

In India's Northeastern borderlands, debates and discourse on indigeneity have been hyper-politicized in terms of human–human relations. As such, the focus has been predominantly statist (Beteille 1986; Karlsson and Subba 2006; Shneiderman and Turin 2006; Xaxa 1999). This body of scholarship reflects the overriding concern with political indigeneity, but at the expense of ecological indigeneity. It downplays indigenous interspecies relations, including their impact on the environment (Todd 2018; Karlsson 2018).

Moving beyond such conceptualizations premised on human political terms, I propose to understand indigeneity through the granularity of more-than-human relations by connecting paddy to the earth, other nonhumans, and humans in ecological and climatic terms. I build on Smyer Yü's (2023, 19) explanation that "being indigenous" always involves "more-than-human experiences in geographically and ecologically specific places, in which human and nonhuman spin together their shared but negotiated and contended life-worlds." Among the Naga, and the soil that nourishes them, worlds are made around paddy's nourishing presence, and increasingly unmade and remade under the influence of anthropogenic transformations.

Many indigenous scholars in India's Northeast reference the United Nations' (2004, 2) definition of the term, which holds that indigenous peoples are "those which, having a historical continuity with pre-invasion and pre-colonial societies that developed on their territories, consider themselves distinct from other sectors of the societies now prevailing on those territories, or parts of them." To some extent, the same can be said of indigenous plants and animals. However, a relational definition vis-à-vis their symbiotic indigeneity is arguably more pertinent. Conceptualizing through the New Himalayas, Smyer Yü (2021, 246) suggests understanding symbiotic indigeneity as follows:

> Indigeneity in the symbiotic sense is a human ecological universal and its varieties are deeply entangled together, like a body of rhizomes, in the New Himalayas. Here, indigeneity and exogeneity are relative concepts. When both are in a symbiotic relationship, they are symbionts in a relational ontology simultaneously involving dependence, interdependence, hybridity and co-becoming for identified common purposes.

This definition stretches beyond the bios—that is, the perceived living organisms that are listed in science textbooks—and acknowledges the possibility of new integral connections and collaborations, not least across political boundaries and multidimensional space. It also restricts the human tendency of exclusive claims to land and territories. Relatedly, I define indigenous crops as those that have been well integrated into the local ecosystem and evoke affinity in their producers and consumers—here, the indigenous people. As such, indigenous crops have moral–political functions and form part of intersubjective multispecies networks.

In Naga cosmology—which is akin to those of many other indigenous peoples across the world—humans are just one of many elemental entities constituting the social world. In that sense, indigeneity is relevant not just for human beings but for myriad other lifeforms that inhabit a particular place as well. For any Naga, the forest, hills, rivers, and rocks are not just quiescent ecology. Rather, they are agential beings and cohabiters with a multiplicity of meanings.

Nagas have culturally specific images, knowledge, and concepts relating to these physical landscapes that affect how they interact with them. The forest, for instance, is an extension of the Nagas' social world, characterized by the presence of spirits that influence their everyday lives. In the not-so-distant past, every tribe actively acknowledged this presence through what they did and what they refrained from doing. Stories are still often told of these spirits, which in turn continue to shape community rules governing behavior toward the forest—what Kire (2014) calls "forest etiquette." In her book *When the River Sleeps*, Kire (2014) introduces two characters: Vilie and an angry forest spirit. When Vilie suffers an injury, he instantly knows that the spirit has punished him because he failed to abide by the "rules of hospitality" his mother had taught him. Through such stories, Kire and several other Naga authors have been key disseminators of Naga lifeways and lifeworlds, especially those pertaining to more-than-humans (Wijunamai 2023).

However, this raises the question of why today's Nagas need to rely on novelists to transmit such knowledge. Perhaps it is because, with Christian conversion, we have experienced a progressive narrowing of our social world until it has become exclusively human, at least in public discourse. On the ground, in the (paddy) fields, farmers and elders with emplaced and embodied experiences still talk of feeling the presence of other-than-humans, but they are silenced at the hearth, where culture and knowledge are habitually transmitted. Nonetheless, the farmers continue to relate to paddy as beings rather than commodities (Banerjee and Wouters 2022). Similar to the gestalt in other societies of highland Asia, in the Naga uplands "[rice] grain lives a parallel life of its own … [It is] a mediator of life and death and is what supports the continuity of society" (Aijmer 2017, 73). This persistence of traditional beliefs is not unique to the Nagas, of course. Writing about the Garos, the anthropologist Erik de Maaker (2013, 159) observed:

> Conversion to Christianity did not result, so far, in the eradication of all traditional divinities. Some are believed to have "gone silent," and are contained, as it were, by the Christian God. Other "lower gods," however—the more malicious ones—have not gone silent at all; rather, they continue to inflict illness and deprive people of their life fluid.

While traditional subjectivities persist in the Naga uplands, recurrent agrarian stress induced by global climate change has increasingly kept people aloof from such more-than-human interactions. In this way, climate change is concurrent with the aforementioned narrowing of the social world.

Conclusion: Anthropogenic Transformations and the Nagas' Paddy Clime

In the Tenyimi Naga tradition, the agro-ecological cycle is variously discussed with reference to typologies of rain. As such, our symbiotic engagement with coexistent lifeforms, including paddy, is determined by the type of rain. Rain determines our food security and our ability to work, but most importantly it is the key to nourishing and maintaining the relationship between Nagas and paddy. But rain is also providence, and the medium through which God blesses the communion of humans and paddy. The following is a prayer for rain, in Chokri (a Tenyidie dialect) by a Christian priest in Thevopisü Village, Nagaland:

> Let the rain that is sweet fall
> Let medium rain fall
> Not the rain that is too strong
> So that it breaks the young rice
> Not the rain that is not enough
> But give us the medium rain
> That will help the grain to grow
> Let the rain that is sweet fall.
> (Translated by Kire 2017b)

Tenyimi ancestors invoked "good rains" for their paddies by performing rituals to appease the gods. The *kemovo* (in Angami)—the village priest—performed a ritual called *tei chü* (lit., "sky-making" or "climate-making") just before the harvest season. On the day of the ceremony, villagers observed *genna*—abstention from work. As part of the ritual, an unblemished heifer was tied to Ora Tobo, a monolith in Mekhroma Village, in present-day Manipur. The *kemovo* then observed and interpreted the animal's expressions and movements, from which he forecast the upcoming weather. If the rope snapped during the ceremony, it was believed that "contact" with the sky could not be made. Villagers took this as a bad omen and prepared for a poor harvest.

Rain also determined the nature of politics in the Naga uplands. During the colonial era, for instance, the Nagas were "independent" during the rainy season, as no expeditions were possible, but "controlled" during the dry season. Hence, the timing and type of rain had a profound impact on the politics, ecology, and sociology of Naga life.

Tenyimi Nagas identify and name rains according to both their time of arrival and their nature. Based on her mother's indigenous knowledge, Pfokrelo Kapesa (2020), a Mao Naga researcher, writes that the first rain in the Mao vernacular lunar calendar is the *okhe raho vu*—"the rain that brings animals and birds." This rain inaugurates life on earth. The weather is warmer, birds start singing, bugs start humming, and animals become active. This is followed by the arrival of the pre-monsoon rain—the *chaka chüshü sowe*—which matures the prevernal crops. Next, around the middle of the year, comes the *kape chiirii*, which is essential for the sowing of paddy. Finally, there is the *ochü mohrükoso*, which comes in two waves—the *ochü mohrükoso karina* and the *ochü mohrükoso konoma*—and continues to fall until the paddy harvest.

In other Tenyimi dialects, the names of various types of rain refer to their nature and magnitude rather than the season in which they appear. For instance, Kuolie, of the Tenyidie Department at Nagaland University, mentioned *rükhrie* (windswept rain), *teise* (rain that is accompanied by thunder), *teimela* (rain that falls with lightning), *teirü toutou* (heavy rain), and *sezhuo* (continuous rain).

Today, in the Naga uplands, clime change is caused by rains becoming ever more erratic, much to the frustration of the farmers, who are finding it increasingly difficult to identify and anticipate rain for their paddy crops. Climate change is locally felt, experienced, and communicated among community members through rice and rain talks. Historically, rice paddy needed the help of humans to become dominant in the Naga uplands, but this process was facilitated by "good rains." Thereafter, both humans and paddy became symbiotically indigenous to this part of the world. Now, though, paddy is losing its dominance. Endorsed by the state, large tracts of land, including paddy fields, are being converted to plantations for areca nut, rubber, oil palm, and horticulture crops, on the pretext that paddy cultivation is no longer sustainable.

Agricultural officers told me that rice yields have declined significantly owing to deficient monsoon rains and "drought-like" conditions. Even if Naga farmers still choose to grow rice, their harvests are poor and/or the paddy is not as healthy as it once was. Exacerbating these problems is paddy's increasing susceptibility to disease due to climate-change-induced biotic stress. In recent years, *Pyricularia oryzae* (popularly known as "rice blast fungus") has damaged a substantial proportion of the standing crops in many parts of the Naga uplands. While indigenous varieties continue to be more resilient than so-called high-yielding varieties (HYVs), plant pathologists suggest that changes in climatic conditions—by way of changing patterns in pathogen dissemination and development rates—have made all rice more susceptible to new diseases (Bevitori and Ghini 2014). Taking all this into consideration, it has become structurally vulnerable and precarious for paddy to grow and thrive in the Naga uplands. Climate change, accordingly, is measured by changes in the quantity and quality of indigenous rice that the Nagas are able to grow.

At the start of this chapter, I mentioned that my mother was delighted to receive *meusan* as a gift. The fact that this is now an increasingly rare and special gesture reflects the extent of anthropogenic transformation in the Naga uplands.

Notes

1 Unless otherwise mentioned, all names used in this chapter are pseudonyms.
2 For an introduction to "plant turn," see Sheridan (2016).
3 Rob Nixon (2011, 4) defines "slow violence" as "incremental and accretive, its calamitous repercussions playing out across a range of temporal scales."
4 The twelve tribes are: Angami, Chakhesang, Pochury, Rengma, Zeme, Liangmai, Rongmei, Inpui, Mao, Poumai, Tangal, and Maram.
5 Zeliangrong (now increasingly referred to as Zeliangrong-Inpui) constitute four Tenyimi tribes: Zeme, Liangmai, Rongmei, and Inpui (also called Puimei). They share a common ancestry through Makuilongdi Village in present-day Manipur.

Acknowledgments

Much of the data for this chapter was collected through friends and relatives. I would like to thank Gairan, Meno, Avinuo, Sengmei, and Easterine for their help and support. I am also indebted to Jelle for thinking with me and for the generous editorial assistance that helped shape the chapter.

References

Aijmer, Göran. 2017. "Rice in the Imaginary World of the Western Angami Naga: Cultural Semantics and Cultural Modalities." *Journal of Ritual Studies* 31(1): 63–75.

Banerjee, Milinda and Jelle J.P. Wouters. 2022. *Subaltern Studies 2.0: Being against the Capitalocene*. Chicago: Prickly Paradigm Press.

Baxter, Alan T., Jonathan C. Aitchison, Sergey V. Zyabrev, and Jason R. Ali. 2011. "Upper Jurassic Radiolarians from the Naga Ophiolite, Nagaland, Northeast India." *Gondwana Research* 20(2–3): 638–644.

Besky, Sarah. 2014. *The Darjeeling Distinction: Labor and Justice on Fair-Trade Tea Plantations in India*. Berkeley, Los Angeles, and London: University of California Press.

Beteille, Andre. 1986. "The Concept of Tribe with Special Reference to India." *European Journal of Sociology/Archives Européennes de Sociologie* 27(2): 296–318.

Bevitori, Rosangela and Raquel Ghini. 2014. "Rice Blast Disease in Climate Change Times." *Rice Research: Open Access* 3: e111.

Callaway, Ewen. 2014. "Domestication: The Birth of Rice." *Nature* 514(7524): S58–S59.

Callison, Candis. 2014. *How Climate Change Comes to Matter: The Communal Life of Facts*. Durham, NC: Duke University Press.

Carey, Mark and Philip Garone. 2014. "Forum Introduction." *Environmental History* 19(2): 282–293.

Carney, Judith A. 2001. "African Rice in the Columbian Exchange." *Journal of African History* 42(3): 377–396.

Choudhury, Baharul, Mohamed Latif Khan, and Selvadurai Dayanandan. 2013. "Genetic Structure and Diversity of Indigenous Rice (*Oryza sativa*) Varieties in the Eastern Himalayan Region of Northeast India." *SpringerPlus* 2: 1–10.

Das, Debojyoti. 2018. *The Politics of Swidden Farming*. London and New York: Anthem Press.

Das, Debojyoti. 2020. "From Millet to Rice." *Asian Ethnology* 79(2): 377–394.

de Maaker, Erik. 2013. "Have the Mitdes Gone Silent? Conversion, Rhetoric, and the Continuing Importance of the Lower Deities in Northeast India." In *Asia in the*

Making of Christianity, edited by Richard Fox Young and Jonathan A. Seitz, 135–159. Leiden and Boston: Brill.

Dey, Arnab. 2018. *Tea Environments and Plantation Culture: Imperial Disarray in Eastern India*. Cambridge and New York: Cambridge University Press.

Elwin, Verrier, ed. 1969. *The Nagas in the Nineteenth Century*. London: Oxford University Press.

Fleming, James Rodger. 2010. *Fixing the Sky: The Checkered History of Weather and Climate Control*. New York: Columbia University Press.

Fornasiero, Alice, Rod A. Wing, and Pamela Ronald. 2022. "Rice Domestication." *Current Biology* 32(1): R20–R24.

Furnes, Harald, Yildirim Dilek, Guochun Zhao, Inna Safonova, and M. Santosh. 2020. "Geochemical Characterization of Ophiolites in the Alpine–Himalayan Orogenic Belt: Magmatically and Tectonically Diverse Evolution of the Mesozoic Neotethyan Oceanic Crust." *Earth-Science Reviews* 208: 103258.

Government of Nagaland. 2017. *Compendium of Rice Landraces of Nagaland Government of State Agriculture Research Station*. Mokokchung: Department of Agriculture, Government of Nagaland.

Hazarika, Manjil. 2017. *Prehistory and Archaeology of Northeast India Multidisciplinary Investigation in an Archaeological Terra Incognita*. New Delhi: Oxford University Press.

Hazarika, Manjil. 2023. "Domesticating Paddy." In *The Routledge Companion to Northeast India*, edited by Jelle J.P. Wouters and Tanka B. Subba, 138–143. Abingdon: Routledge.

Heneise, Michael. 2018. *Agency and Knowledge in Northeast India: The Life and Landscapes of Dreams*. Abingdon and New York: Routledge.

Huber, Toni and Stuart Blackburn, eds. 2012. *Origins and Migrations in the Extended Eastern Himalayas*. Leiden and Boston: Brill.

Hutton, John Henry. 1921. *The Angami Nagas*. London: Macmillan & Co.

Joshi, Vibha. 2004. "Human and Spiritual Agency in Angami Healing." *Anthropology and Medicine* 11(3): 269–291.

Kapesa, Pfokrelo. 2020. "People who Walks with the Rain." October 8. https://rb.gy/nqekf.

Karlsson, Bengt G. 2018. "After Political Ecology." *Anthropology Today* 34(2): 22–24.

Karlsson, Bengt G. 2022. "The Imperial Weight of Tea: On the Politics of Plants, Plantations and Science." *Geoforum* 130: 105–114.

Karlsson, Bengt G. and Tanka B. Subba, eds. 2006. *Indigeneity in India*. London, New York and Bahrain: Kegan Paul.

Kire, Easterine. 2014. *When the River Sleeps*. New Delhi: Zubaan.

Kire, Easterine. 2017a. *Don't Run, My Love*. New Delhi: Speaking Tiger.

Kire, Easterine. 2017b. "Let the Rain that is Sweet Fall: In Nagaland There Are Many Names for Rain." *Indian Express*, July 9. https://indianexpress.com/article/lifestyle/life-style/let-the-rain-that-is-sweet-fall-in-nagaland-there-are-many-names-for-rain-4741918/.

Lehman, F.K. (Chit Hlaing). 1989. "Internal Inflationary Pressures in the Prestige Economy of the Feast of Merit Complex: The Chin and Kachin Cases from Upper Burma." In *Ritual, Power and Economy: Upland–Lowland Contrasts in Mainland Southeast Asia*, edited by Susan D. Russell, 89–102. DeKalb: Center for Southeast Asian Studies, Northern Illinois University.

Ludden, David. 2005. "Where is Assam?" *Himal Southasian*, November 15. www.himalmag.com/where-is-assam/.

Mandal, Jaydev and T.R. Shankar Raman. 2016. "Shifting Agriculture Supports More Tropical Forest Birds than Oil Palm or Teak Plantations in Mizoram, Northeast India." *The Condor: Ornithological Applications* 118(2): 345–359.

Mayirnao, Shaokhai and Sinalei Khayi. 2023. "Decolonising Feasts of Merit: Reasoning Marān Kasā from a Tangkhul Naga Perspective." *Asian Ethnicity* 24(2): 258–277.

Mills, James Philip. 1935. "The Effect of Ritual upon Industries and Arts in the Naga Hills." *Man* 35: 132–135.

Nixon, Rob. 2011. *Slow Violence and the Environmentalism of the Poor*. Cambridge, MA: Harvard University Press.

Okoshi, Masako, Tomotaro Nishikawa, Hiromori Akagi, and Tatsuhito Fujimura. 2018. "Genetic Diversity of Cultivated Rice (*Oryza sativa L.*) and Wild Rice (*Oryza rufipogon Griff.*) in Asia, Especially in Myanmar, as Revealed by Organelle Markers." *Genetic Resources and Crop Evolution* 65: 713–726.

Pachuau, Joy L.K. and Willem van Schendel. 2022. *Entangled Lives: Human–Animal–Plant Histories of the Eastern Himalayan Triangle*. New Delhi: Cambridge University Press.

Saikia, Arupjyoti. 2019. *The Unquiet River: A Biography of the Brahmaputra*. New Delhi: Oxford University Press.

Scott, James C. 2009. *The Art of Not Being Governed*. New Haven: Yale University Press.

Sharma, Jayeeta. 2011. *Empire's Garden: Assam and the Making of India*. Durham, NC, and London: Duke University Press.

Sheridan, Michael. 2016. "Boundary Plants, the Social Production of Space, and Vegetative Agency in Agrarian Societies." *Environment and Society* 7(1): 29–49.

Shneiderman, Sara and Mark Turin. 2006. "Seeking the Tribe: Ethnopolitics in Darjeeling and Sikkim." *Himal Southasian* 18(5): 54–58.

Smyer Yü, Dan. 2021. "Symbiotic Indigeneity and Commoning in the Anthropogenic Himalayas." In *Environmental Humanities in the New Himalayas*, edited by Dan Smyer Yü and Erik de Maaker, 239–260. Abingdon: Routledge.

Smyer Yü, Dan. 2023. "Multipolar Clime Studies of the Anthropocenic Himalaya, Andes and Arctic: An Introduction." In *Storying Multipolar Climes of the Himalaya, Andes and Arctic: Anthropocenic Climate and Shapeshifting Watery Lifeworlds*, edited by Dan Smyer Yü and Jelle J.P. Wouters, 1–26. Abingdon and New York: Routledge.

Smyer Yü, Dan and Jelle J.P. Wouters, eds. 2023. *Storying Multipolar Climes of the Himalaya, Andes and Arctic: Anthropocenic Climate and Shapeshifting Watery Lifeworlds*. Abingdon and New York: Routledge.

Stonor, Charles R. 1950. "The Feasts of Merit among the Northern Sangtam Tribe of Assam." *Anthropos* 45(1–3): 1–12.

Todd, Zoe. 2014. "Fish Pluralities: Human–Animal Relations and Sites of Engagement in Paulatuuq, Arctic Canada." *Études/Inuit/Studies* 38(1–2): 217–238.

Todd, Zoe. 2018. "Refracting the State through Human–Fish Relations: Fishing, Indigenous Legal Orders and Colonialism in North/Western Canada." *Decolonization: Indigeneity, Education and Society* 7(1): 60–75.

United Nations. 2004. "The Concept of Indigenous Peoples." January 19–21. www.un.org/esa/socdev/unpfii/documents/workshop_data_background.doc.

van Driem, George. 2017. "The Domestications and the Domesticators of Asian Rice." In *Language Dispersal beyond Farming*, edited by Martine Robbeets and Alexander Savelyev, 183–214. Amsterdam and Philadelphia: John Benjamins Publishing Company.

von Fürer-Haimendorf, Christoph. 1935. *The Naked Nagas.* New York: AMS Press.

Wijunamai, Roderick. 2023. "Understanding Naga Cosmology through Easterine Kire's Fictional Narrative." In *Keeper of Stories: Critical Readings of Easterine Kire's Novels*, edited by K.B. Veio Pou, 29–42. Kohima: Highlander Press.

Wouters, Jelle J.P. 2015. "Feasts of Merit, Election Feasts or No Feasts? On the Politics of Wining and Dining in Nagaland, Northeast India." *South Asianist Journal* 3(2): 5–23.

Wouters, Jelle J.P. 2023. "Multilateral Clime Studies." In *Storying Multipolar Climes of the Himalaya, Andes and Arctic: Anthropocenic Climate and Shapeshifting Watery Lifeworlds*, edited by Dan Smyer Yü and Jelle J.P. Wouters, 53–72. Abingdon and New York: Routledge.

Xaxa, Virginius. 1999. "Tribes as Indigenous People of India." *Economic and Political Weekly* 34(51): 3589–3595.

Yekha-ü and Queenbala Marak. 2021. "Elicüra: The 'Feasts of Merit' Shawl of the Chakhesang Naga of Northeast India." *Oriental Anthropologist* 21(1): 138–157.

3 Lakes in Life

Mermaids and Anthropocenic Waters in the Bhutan Highlands

Jelle J.P. Wouters and Thinley Dema

Introduction

From the annals of Deep Time to the Anthropocenic present, the Himalayas have been intimately bound up with water. The Himalayas can be helpfully framed as a "terrestrial ocean" (Smyer Yü 2021, 9) for their near-unparalleled capacity to store and circulate water (in its liquid, vaporous, and solid phases). What is more, the range's orogenesis was a distinctly watery process. Marine fossils and limestone are routinely located in the high Himalayas, revealing these peaks' and summits' previous geological life as an ocean bed. Indeed, when Tenzing Norgay and Edmund Hillary planted their flags on Sagarmāthā (Nepali)/Chomolungma (Tibetan)/Mount Everest (English) in 1953, they "set them in snow over the skeletons of creatures that had lived in the warm clear ocean that India [in its tectonic form], moving north, blanked out" (McPhee 1999, 124). Read thus, the epic of Himalayan orogenies is, significantly, an oceanic history.

To critically apprehend this mountain–ocean mutuality, we must venture far into the near-mystical world of Deep Time (see Irvine 2020) that relates the agency and architecture of the indigenous and pre-human earth in a series of geological epochs in which everything was possible—from mountain-building and glacier-making to moving valleys and traveling rivers. Himalayan orogenies replaced the ancient Tethys Sea, which was parted and drained by the crumpling, buckling, faulting, uplifting, and folding of the earth's crust as the Indian and Eurasian plates entered into an intra-material vitality, conjuring a "vital materialism" (Bennett 2010), expanded to include tectonic interactions as antecedently forming the geo–eco–climatic basis of human and other-than-human life. The ancient inland sea that once connected the two supercontinents of Laurasia and Gondwana, and of which remnants and traces are everywhere present in the surface and subsurface of the Himalayas, was given the name Tethys in 1839 by the Austrian geologist Eduard Suess in honor of the Greek sea-goddess, a Titaness who was Gaia's daughter, wife of Oceanus, and mother of rivers, lakes, and clouds. Tethys, then, in its indigenous Greek association, resonates with diverse indigenous Himalayan conceptualizations that affirm the integral union of watery,

DOI: 10.4324/9781003484394-3

affective, spiritual, and political worlds. It surely partakes of the ontological status of bodies of water in the Bhutan highlands, where water, as this chapter will show, is always also more-than-water.

Offering an account of water, this chapter climes lakes in the Bhutan highlands, particularly the Thimphu highlands that border the Tibetan Autonomous Region to the north.[1] Our overall understanding of "clime" is that of a multispecies dwelling place embodied with climate history, patterns, and changes; thus, "lake climes" are places where lakes are central to ecological and cultural communities of life. In this chapter, however, we will use "clime" predominantly as a verb, as in "climing" (Paerregaard 2023; Smyer Yü 2023), in order to attune to the complex, multiply-mediated processes, deeply tied to emplaced imaginations, through which humans and other-than-humans emerge and merge with the earth's life-enabling and sustaining material affordances as they move and change across the Critical Zone (see Smyer Yü 2020), paying particular attention to waters in their embodied lake form. Specifically, climing lakes entails attentiveness to the eco-material, affective, experiential, somatic, spiritual, and overall relational and agential entanglements between humans and lakes, including the interpretations and navigations of—and responses to—ongoing anthropogenic transformations of lakes. Situating ourselves in the emergent "blue humanities" (Mentz 2023), "hydro-humanities" (De Wolf, Faletti, and López-Calvo 2021), and more-than-human humanities more generally, we show how agential and affective entanglements between lakes, humans, and other-than-humans, grounded in clime studies, can expand and multiply our understanding of water in an Anthropocene context. Such lively perspectives complement (and potentially transform) scientific knowledge-making and promote conversations about water that are more capacious, caring, curious, critical, and inclusive than those dominated—in Bhutan, as elsewhere—by scientists, engineers, hydrologists, economists, and politicians.

Water, we know, is the earth's most common environmental flow. Water, we also know, is life. It is ubiquitous on, above, and beneath the earth's surfaces. Moreover, in many Himalayan settings, water fully immerses such surfaces, as much as it fills and fuels the cells of living beings. Global waters everywhere and always exist in relation to other-than-waters, whether of bodily, other material, or discursive substance. In the relations they garner, waters are revealed as makers, movers, and changers of social, political, ecological, climatic, and spiritual worlds. While they are everywhere life-giving and world-making, they are simultaneously place-specific and diversely acknowledged to have multiple ontological statuses and capacities of action and affect across material, cultural, political, spiritual, and symbolic domains that vary between epistemes and spatio-temporal contexts.

In watery terms, the Anthropocene is unique due to persistent attempts, orchestrated by state capital, to make waters "modern," in the sense, as Jamie Linton (2010) explains, of an anthropogenic universalizing, homogenizing, and reducing of water to mere H_2O by depriving it of its plural ontological realities. These modern—or anthropogenic—waters are variously managed, contained, channeled, commoditized, polluted, diverted, and dammed by anthropogenic structures, activities, and desires. Of course, like humans (Latour 1991), these supposedly modern waters have never been truly modern as they regularly refuse to passively obey human will. Instead, they retain their power and agency to disrupt human (and other) lives and anthropogenic structures through life-taking excesses (floods) and absences (droughts). More than this, Anthropocenic waters, in their various flows, currents, and bodies, increasingly shapeshift and migrate so that, from a human perspective, they are often in the wrong phase and/or the wrong place.

Amid global anthropocentric transformations of water, this chapter attunes to lakes *who* (not which) are habitually named and personified in the Bhutan highlands, who manifest as bodies of spiritual instruction and enchantment, who are revealed as active protagonists in shared more-than-human worlds, and who, therefore, are always "more-than-water." Through ethnography and experience narratives, we show how highland herders clime lakes as they socially, spiritually, and ritually engage with lake-bodies, whom they regard as inhabited by numinous *tshomen* (mermaids).[2] In their climing practices, the herders envision and enact a shared more-than-human world attuned and attentive to the agency, history, subjectivity, and anthropogenic changes of/in water bodies/beings.

Bhutan as an Aquatic Land

Before introducing the main protagonists of this chapter—namely, herders and *tshomen*-lakes—Bhutan's physical characteristics, and its water tables and stresses, will be briefly noted. Bhutan's land is distinctly vertical, with subtropical lowlands in the south ascending into temperate midlands that climb into a northern belt of alpine highlands that are walked by yaks and herders. The terrain reaches 7578 meters with the peak of Gangkhar Puensum, which translates as "White Peak of the Three Spiritual Brothers." This peak is too sacred to scale, making it the world's highest mountain that is protected from human climbing. Bhutan's distinctly vertical geomorphology voices a complexly layered history of space and temporality that began roughly 50 million years ago with the Himalayan uplift (see Pandit 2017). This earth epic of mountain-making was simultaneously a process of climate-making (see Chapter 1, this volume), and locally the co-maker of glaciers, snowscapes, permafrost, rivers, and lakes.

At high altitudes, these earth creations mutually constitute Bhutan's cryosphere, which is an integral part of the pre-human Himalayan cryosphere. Exactly where Bhutan's cryosphere begins and ends is hard to define, first because of problems of definition (e.g., measurements of the glacial influence on lakes and frost action) and second because the cryosphere is continuously on the move through variegated melt and growth phases (Sörlin 2015). However, it is generally held that roughly 10 percent of Bhutan's landmass is part of the cryosphere (Cryosphere Services Division, National Center for Hydrology and Meteorology 2021). Bhutan's cryosphere is integral to an intricate vertical drainage network that co-feeds an archipelago of lakes and rivers (Wang Chhu, Punatsang Chhu, Manas Chhu, Amo Chhu, Nyere Ama Chhu, among others). These rivers become emboldened as they funnel down into nearly 200 watersheds, and as trees and shrubs take up water through their roots and slow down the runoff while increasing soil drainage. In their catchments, these watersheds are unparalleled in the Himalayas and form out of Bhutan's extensive forest cover (over 70 percent of its landmass). As a result, water in Bhutan is stored everywhere in the landscape and recharges surface-water bodies and underground aquifers. This renders Bhutan an "aquatic land" or "terrestrial ocean." We borrow these terms from Dan Smyer Yü (2021, 9), who propounds them for the wider Himalayas in full recognition of the immensity of water in both solid and liquid forms throughout in the landscape. Figure 3.1 relates this ubiquity of water and reveals its performativity as it inundates land.

Current climate research in Bhutan strongly focuses on changes and stresses in its water tables and cycles. Glaciologists, theoretical meteorologists, atmospheric chemists, and modelers document with increasing alarm the anthropogenic

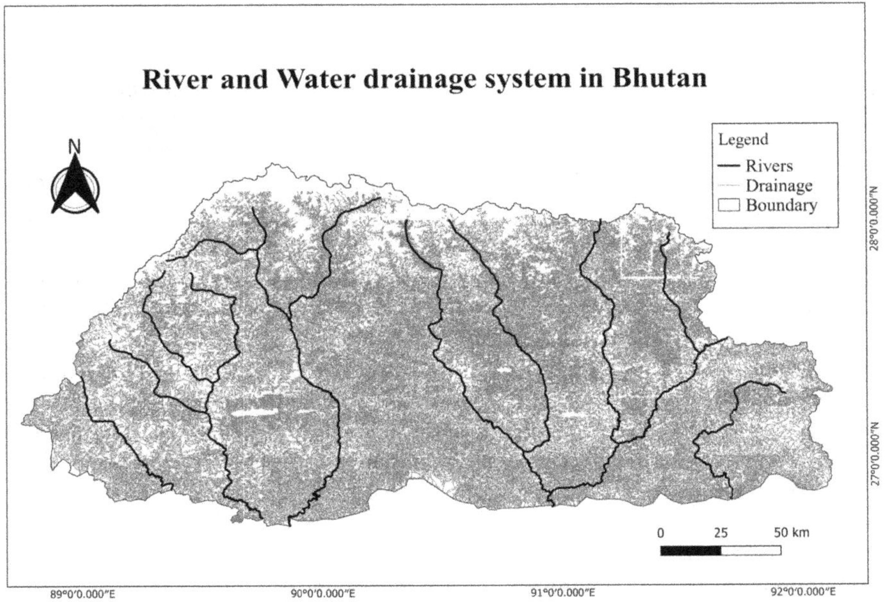

Figure 3.1 Map of the drainage system and rivers in Bhutan
Note: There is less drainage in the north due to permanent ice cover. The flow pattern is distinctly north–south. The fact that almost the entire map is shaded indicates the abundance of water on, below, and intermeshed with the earth's surface.
Source: The authors

interpellations among the atmosphere, cryosphere, hydrosphere, and biosphere, and how these impact glacial and highland environments in Bhutan and beyond (Bajracharya et al. 2014; Kumar et al. 2021; Wangchuk et al. 2022). Highland lakes are at the forefront of this scientific climate research because they are acknowledged as barometers of local climatic conditions and changes therein, and because of Bhutan's historical and modeled future of devastating glacial lake outburst floods (GLOFs), which result from excessive precipitation and accelerated glacial and snow melt. Ever since a destructive lake outburst killed 21 people in 1994, Bhutan has surveyed over 2600 glacial lakes (Cryosphere Services Division, National Center for Hydrology and Meteorology 2021; Tshewang et al. 2021). Twenty-four of these are classified as immediately dangerous to humans (and other-than-humans), as excessive precipitation and accelerated snow melt are rapidly adding water to them, with potential for a "tsunami in the sky" (Cryosphere Services Division, National Center for Hydrology and Meteorology 2021). These lakes are now constantly monitored, and their water levels are balanced by channeling water out of them. Elsewhere in the Bhutan highlands, because of micro-climatic variabilities, the ecological and cultural problem is not lakes overflowing but of them drying up and disappearing as a result of fluctuating precipitation patterns, increased evaporation due to temperature rise, the vanishing of glaciers, and an overall reduction in the mountains' and highlands' capacity to circulate and store water.

Whether in relation to floods or droughts, the prevailing science-oriented scholarship, and the policy interventions it informs, generally apprehends lakes as physical properties that are fully comprehended by physics and chemistry. While these researches are vitally important, they ipso facto essentialize highland lakes as "only" objects of scientific inquiry, and in their knowledge-making they generally distance, detach, and disembody lakes from humans, nonhumans, and the ecologies of relations that intermesh them, except in the broad sense of global anthropogenic forces that impel local hydraulic, ecological, and climatic changes. As this chapter will show, highland herders live with a more complex, experiential, and layered understanding of the lakes who variously interpolate into their lives as relational and inter-agentive beings. Herders know that lakes are also subjects with life-histories and other-than-human modes of agency, perception, communication, and temporality, and their everyday climing practices reflect this understanding of lakes as relationally alive. Thus, a focus on highland herders' knowledge, experiences, and climing practices complements climate science by widening the reach of lake agency and relationality beyond the laws of pure physics and chemistry. Even as ever more research attention and funding are allocated to scientifically "reading" changes and transformations in highland lakes in Bhutan and across the Himalayas, the highland herders of this chapter know that lakes also "read back" and respond to changes in human activities. To put this in clime terms, as much as herders are climing lakes, the lakes are also climing herders.

The next section offers an ethnographic vignette to set the scene, context, and narrative argument of this chapter.

Lake-Yaks

Jigme is a yak-herder who co-dwells and co-migrates with his animals in the Bhutan highlands. He knows the landscape and witnesses changes, however minute, in the density of the grasses his yaks graze, in the silhouettes of snowlines and glaciers, in the intensity of rains, and in the color of lakes—of which there are many. Some of these lakes, both large and small, are fed by glacial melt; others are nourished solely by groundwater and precipitation.

Jigme herds well over a hundred yaks, yet he has an intimate connection with them, knowing each one by name, genealogy, and personality (Wouters 2021). Three of the animals are particularly precious to him because they belong to the *tshoyak* (lake-owner) lineage. Their appearance sets them apart from their peers: they have milky eyes, comparatively large bodies, and thick, hard, sharp fur—"like needles," as Jigme puts it. Their *apa* (father)—or *zhuel*—lives with/in the *tsho* (lake), who in this case, and generally, is also a *tshomen* (mermaid). To bestow *ngodrup* (blessings) on Jigme, the *tshomen* sent a male yak out of the *tsho* to mate with the herder's "land-yaks." The *zhuel* then retreated back into the lake. The progeny that resulted from this more-than-yak encounter stand stout and proud, reflecting their superiority over the other members of the herd. They provide both status and spiritual value to Jigme, who is prohibited from killing or selling them as they are "lake blessings."

Jigme, indeed, considers himself blessed. He vividly recalls an incident that occurred some years ago. Sitting near the lake that is close to his *tsamdro* (pasture), he saw a white *dunkhar* (conch shell), symbolizing the reverberating sound of the Buddha Dharma, spinning around the water. Recognizing the *dunkhar* as one of the *Tashi Tagye*—the eight auspicious symbols of Buddhism—he immediately interpreted it as a message from the lake-owner, the *tshomen*, signaling her favor. Ever since that day, he has conscientiously climed the lake by offering milk and incense to the *tshomen* whenever he passes by. In addition, he has ensured that he and his relatives do not introduce any *drib* (variously pollution, profanation, contamination, obstruction, or negative energy) to the lake, for instance by talking loudly, defecating, burning meat or chilies (the smell of which the *tshomen* is known to dislike), or thinking "impure" thoughts. *Drib* is a key force in Bon and Tibetan Buddhism that can take on physical, social, and cognitive forms. Toni Huber (1999, 16) explains:

> In common thinking *drib* is generally conceived as a form of both physical and social pollution that is associated with various substances and proscribed social practices and relations, as well as with deities inhabiting both the body and the external world.

Drib also takes an ethical–spiritual form in its association with the *dikpa* (sin) that a person accumulates and bears. *Drib-dik* are generally viewed as "negative, obstructive, unlucky, and even threatening (to health, longevity, fertility, prosperity, etc.) aspects of ordinary human social and material

Figure 3.2 A *tshoyak*
Source: Photo by Pema Choden, 2022

existence" (Huber 1999, 16). Highland herders in Bhutan regularly mention *drib* when discussing lakes, especially in response to experienced anthropogenic alterations in human–lake relations.

Although Jigme cherishes his lake-yaks, he worries about them. Elderly herders have warned him that the animals generally do not live long because their half-spirit origins make them particularly vulnerable to earthly defilement (*tsokpa*), which threatens their health. Jigme's experiences with his lake-yaks and the *dunkhar* emerge from a relational field in which any action is also an interaction, as the whole highland landscape, including mountains, lakes, glaciers, streams, cliffs, and grasslands, is known to be relationally aware and responsive to human (in)action. Here, dialogues and bonding, affinity, and mutuality, as well as fear and awe, routinely move beyond the domain of the human to create and nourish worlds that are shared between various beings—humans and other-than-humans, eco-materialities such as mountains and lakes, and non-secular beings of Bon–Buddhist heritage, such as *tshomen*. While it is certainly possible to feel *lonely* in the highlands—indeed, many herders acknowledge the emotion when co-migrating with their yaks—it is impossible to be *alone*, as each territory has its non-secular owners in what is understood as a cosmic polity that hierarchically encompasses humans (Wouters 2023).

In terms of highland lakes, herders and *tshomen* cultivate real and viable relationships that are alive and consequential. Our ethnography and narratives compel us to argue that in the Bhutan highlands, as much as life is in lakes, because lakes are integrally life-giving by nourishing biomes and beings, lakes are also in life. By this, we imply the consequential agential entanglements that connect humans and *tshomen*-lakes. As preceding narrative has demonstrated, and as we unpack further below, in the Bhutan highlands, the importance of lakes goes well beyond their eco-material presence. In what follows, we show that *tshomen*-lakes listen, act, create, respond, and disturb in ways that are generally considered exclusively human. In this way, they actively co-participate in the production of the social, material, ecological, and spiritual worlds they inhabit alongside humans and other-than-humans.

Cosmological Introductions: Who Are the *Tshomen*?

In their bodily appearance, *tshomen* are encountered and realized by highland herders as part female and part reptilian: that is, similar to the universalist mermaid motif that features prominently in folklore from Africa to Asia, Europe to South America. In this lore, mermaids variously manifest as benevolent, malevolent, irresistible, and the cause of floods, storms, shipwrecks, and drowning. *Tshomen* do not always appear in the physical embodiment of a mermaid, however. Herders know that they may also shapeshift into yaks, horses, octopuses, fish, snakes, or humans. While numinous beings are notoriously hard to trace, track, and theorize, they are unmistakable indigenes—the original occupants of Bhutan's terraqueous habitat. This makes them members of a broader category of animate beings whom Karma Ura (2001, 1) characterizes as "immortal owners." He asserts that Bhutan is the

> domain of occupants other than plants, animals, and human beings. There are other beings to whom geo-sensitive areas of our country have been ascribed, and to whom, beyond human ownership, they belong. *Nas dag, zhi dag*, and *yulha* [various types of deity] are immortal owners or landlords, while successive generations of communities are ephemeral travelers passing through their territory.

Within this relational environment, "calm lakes or deep parts of the rivers are homes of *Tsho man* [tshomen] (lake women)" (Ura 2001, 5). *Tshomen* also inhabit other terraqueous areas, such as meadows and marshlands. The physical presence of lakes, rivers, and marshlands was antecedent to human arrival, and there was no pre-established harmony between *tshomen*, other numinous beings, and humans. Indeed, the land was intransigent and hostile to human settlement. To offset this human–nonhuman antagonism, Padmasambhava (also known by his followers as Guru Rinpoche or the Second Buddha) arrived in the eighth century CE and leveraged his spiritual power to broker a covenant between autochthonous deities, including *tshomen*, and humans. The resultant more-than-human social contract became constitutive, in ethnographic terms, of shared norm-worlds that rendered the "immortal owners" of the land receptive and responsive to human presence. It was this that allowed humans to settle, cultivate, and flourish in a land that was otherwise owned by deities. A herder named Dorji provides an insight into the nature of this relationship:

> There are palaces under the lakes where *tshomen* live, and there are reasons why the lakes exist in particular places. Each lake in our highlands has its own story and reason to be there. If the lake dries up or overflows, it is because the *tshomen* is offended. If we herders treat the lake with reverence and do our rituals properly, then the *tshomen* will be pleased and give us *ngodrup.*

Padmasambhava's intervention was deeply transformative for the *tshomen*. It bound them under sacred oath as protectors of the Buddha Dharma (Phuntsho 2013) and charged them with safeguarding *terma* (treasures, revealed knowledge, of which Jigme's *dunkhar* was one manifestation) in their waters.

The sanctity of lakes in the region has prompted the local herders to put their number at 108. This figure symbolizes *kangyur*—the Tibetan Buddhist canon, which is loosely defined as consisting of 108 volumes of sacred texts. In strictly genealogical terms, *tshomen* emerged out of Bon, often considered as Bhutan's original or oldest religion, which then merged with Buddhism. However, in ethnographic terms, Bon and Buddhism mix and blend throughout Bhutan (Pommaret 2014), even as, in terms of philosophy and formal practice, the two also compete and contest (Tashi 2023). The intimate association of *tshomen* with water partakes of broader Bon–Buddhist epistemologies in which water takes on various sacred–sentient significances, whether in terms of *drupchu* (healing/holy water), *lu, lubuum*, and *lusuud* (respectively Nagas, in the sense of numinous serpents associated with watery surfaces, serpent dwellings, and offerings to them), in daily water offerings of seven cups on altars in homes and monasteries, or in prayer verses. Moreover, human existence itself emerges from the elemental confluence of earth, air, fire, and water (a combination known as *jungwa zhi gi ken*).

Crucial to the genealogy of *tshomen* is that Padmasambhava, while transforming their being, also affirmed their status as indigenes with privileged access to and ownership of water, and thus also changed Buddhism by adapting it to local forces. In this way, *tshomen* were transmitted into the Buddhist epoch as autochthonous authorities capable of asserting themselves around, in, and through humans. Recognized as more powerful than humans, they are not merely to be respected or avoided. Rather, humans must monitor them closely because human well-being and welfare significantly depend on adhering to their prescriptions and proscriptions (and those of other terrestrial deities). Moreover, this obligation has to be fulfilled with particular care when dealing with *tshomen* as they are especially sensitive and vulnerable to human (in)action and (im)propriety (Ura 2001).

Simultaneously, *tshomen*'s close association with water relegates them to the lower realms of Buddhist cosmology, as unenlightened beings. For all their power and privilege as first settlers, this renders them envious of humans, who, while similarly unenlightened, possess souls that can be cultivated to allow an escape from *samsara*—the relentless cycle of birth, death, and rebirth. In this way, *tshomen*'s conversion by Padmasambhava also manifests to them as a curse, forever tying them to earthly bodies of water. It is this predicament that feeds speculation of *tshomen* taking on human avatars to attend annual religious *tshechu* (rituals), where they observe masked dances (*cham*) that bestow merit on the onlookers and prepare them for their journey into the afterlife. Because *tshomen* are immortal, the only way in which they can glimpse the afterlife is during one of these *cham* performances. Aum Yudey, a highland herder, explains: "*Tshomen* can bless or harm us while in this world. But as they are tied down to the *samsaric* world, they have no power over us after we die." As original landowners and unenlightened immortals, *tshomen* are thus at once superior and subordinate to humans, which translates in the herders' epistemological reflex of relating the variously benevolent, malicious, covetous, wealth-generating and wealth-extracting, life-giving and life-taking actions of *tshomen* to their agential capacities.

Tshomen, Drib, and Weather Changes

"I hope that outsiders [non-highlanders] will not visit the *tsho* [a two-hour walk from the highland hamlet of Soe] tomorrow," Tandin, a third-grade student remarked. The annual school *rimdro* (ritual) was due to be held the next day, and Tandin was excited by the prospect of the feast and festivities. However, as it turned out, bad weather disrupted the arrangements, including Tandin's plan to spend the post-ritual hours playing with her friends.

Although not yet ten years of age, Tandin—like all her fellow villagers—knows that the lake is the abode of a *tshomen* who is attentive to the presence of humans on the banks. If *drib* is imparted, whether wittingly or unwittingly, the *tshomen* might direct her anger not only to the individual offenders—by causing them physical, mental, or spiritual harm—but to the whole community. After all, she has entered into a covenant with all of them. Like most deities in Bhutan, *tshomen* are zealously territorial in terms of residence and the control, power, ritual, and kinship they cultivate with humans (see Pommaret 2014). They employ various extra-linguistic forms of communication to make themselves known. Prime among these is their impelling of abrupt and volatile weather changes (thick fog, torrential rain, gales), which make them co-makers of the Bhutan highland clime.

In recent months, several non-highlanders, including government officials, military men, and tourists, had visited the lake. A series of sudden, unusual weather events had followed, leaving the herders in no doubt that the spike in human presence and activities had agitated the resident *tshomen*. While the herders have an intimate connection with the *tshomen*, conforming their climing practices to them and understanding their likes, dislikes, and taboos, the same depth of knowledge and consideration cannot be expected of outsiders. Another risk is that outsiders usually travel long distances before arriving at the lake, so they are likely to encounter various beings, both human and numinous, along the way, from whom *drib* and possible demonic substances may rub off. In addition, the herders recognize that *tshomen*—much like themselves—can be moody and fiery, as well as loving and generous. Indeed, it is often said that "*Lha drey mi sum choelam chi* [Gods, ghosts, and humans all have the same behavior]."

The day of the ritual dawned bright and sunny. Tandin happily frolicked with her friends near a gently rippling stream. Meanwhile, Dawa, an elderly herder, was optimistic that everything would run smoothly. Pointing toward a black crow, he said, "See, it is facing the river. Aup Chundu's [territorial deity] cave is just above the river there. It means that Aup Chundu is pleased with our ritual today."

Young Tandin's intuitive remark connecting the lake, *tshomen*, human (in)action, and weather exemplified a much broader, ongoing, more-than-human conversation and commensality in the highlands in which *tshomen*-lakes (but also specific peaks, caves, and cliffs) are agentive actants. A basic fact of anthropology holds that individuals, as social beings, do not pre-exist their interactions but emerge through and as part of entangled intra-relatings. This logic extends to human–lake relations. Herders generally do not construe lakes as a subjective force that emanates from a conscious lake-self, in the sense of acting in isolation and independently. Rather, they conceive of them as fundamental, active participants in the mesh of an all-encompassing relational field. It is precisely in this manner that lakes reveal themselves to be *in* life. In this relational setting, the primarily ontological unit of climing is not the lake independently, but the agential relatings between herders and lakes, as enlivened by *tshomen*.

Besides *ngodrup*, as in the case of Jigme's lake-yaks and spiritual encounter, the notion of *drib* is often invoked to explain human–lake interactions. A herder named Sonam insisted:

> If a person takes the path that goes along the lake up there [pointing northwards], and if he carries *drib*, then the *tshomen* will be offended, the clouds will start forming and it will rain heavily. There might even be hailstorms and thunderstorms.

Figure 3.3(a) Two lakes in the western Bhutan highlands
Source: Photos by Pema Choden, 2022

Therefore, it is along a continuum of relations—from *ngodrup* to *drib*, and often through unexpected weather events—that lakes' agentive forces, propensities, and tendencies, as enlivened by *tshomen*, most forcefully reveal themselves. While *drib* can be caused unconsciously—for instance, through bad karma carried over from a previous lifetime—causing it deliberately can have severe, even lethal, consequences. A Bhutanese film titled *Six Boys* (2003) is based on the real-life story of six schoolboys who disappeared in the forest around Domendrel Tsho, above the capital Thimphu, in 1996. The film shows the boys disrupting the lake by throwing stones into it and making loud noises. In response, the furious *tshomen* summons heavy rainfall and thick fog that make the boys lose their path home. Their disappearance leads to a massive rescue operation, and twelve days later four of them are found—pale, sick, and starving—in distant Punakha. The other two fail to make it out alive.

Figure 3.3(b) (Cont.)

The experiential reality of Bhutan's lakes as makers and changers of weather is, in a very different lexicon and episteme, corroborated by climate science in the understanding of "lake effects" as local and regional fluctuations in temperature, rainfall, snowfall, fog, and wind generated by a body of water's eco-material capacity for absorbing, storing, and moving both heat and liquid (Scott and Huff 1997; Sousounis 2001; Su et al. 2020). For instance, when warm air floats over a colder body of water, the air quickly cools and moistens, causing the dew point to rise. As the dew point approaches the air temperature, condensation occurs and fog droplets form. In the highlands of Bhutan, such droplets frequently coalesce into thick veils of fog that may conceal the lake's physicality altogether. Herders believe that this self-concealment often occurs when strangers approach and the *tshomen*, unsure of their intentions and the *drib* they may carry, raises the fog as a protective curtain.

The above narratives demonstrate that the ways in which herders encounter, understand, and interpret the weather, and fluctuations therein, broadly resonate with the spiritual–moral meteorology that holds sway across the wider Tibetosphere. Whereas, in scientific terms, weather is explained as a system of quantified relationships between a range of atmospheric and terrestrial variables, in the Tibetosphere it also relates "to a system of local qualitative interrelationships of humans and spirit powers" (Huber and Pederson 1997: 577). As we have seen, in Bhutan, this system includes the agential entanglements between humans and *tshomen*. Highland herders know that weather—as well as its systematic extensions: climate and climate change—is not an independent domain that is subject exclusively to the laws of physics and the inert backdrop against which the arbitrariness of human activities enact themselves. Rather, they know that human (in)action (including karmic, moral, ethical, and ritual enactments) is deeply entwined—and often in direct causal conjunction—with fluctuations in weather and climate. In this way, weather (change) is mediated by a complex, multilayered human–nonhuman matrix.

Returning to Tandin's and Dawa's comments, these experiences reveal themselves as the practical embodiments of a deep-seated clime order in which climing humans are environed in a perceiving clime.

Encountering the *Tshomen*: Blessings, Temporalities, and *Jigdra*

Accounts of physical sightings of *tshomen* continue to circulate, despite warnings from elders that herders and outsiders alike have fallen ill, or even died, after boasting of such encounters. In light of this, the frequency of *tshomen* encounters in the highlands might well be higher than we can ascertain from herders' reports.

The encounters can take various forms. As noted earlier, *tshomen* shapeshift, so some herders report seeing them in animal or human manifestations—occasionally in the shape of a relative (either living or dead)—or in the form of a gem, a *dunkhar*, or some other *terma* (treasure). Ashim Dorji recalls:

> It was said that, in the past, fortunate people witnessed a man, a woman, and a yak walking around this lake [*Gay-dhue tsho*]. The man would be wearing a *gho* [Bhutan's male national dress] and holding a rope tied to a yak. The woman would follow him carrying a butter churn on her back.

Today, it is likely that this "lake vision" is invoked to culturally legitimize and frame the gendered division of highland labor: women are mostly engaged in milking the yaks, while men herd them.

Other *tshomen* encounters unfold in dreamtime, when perceptual boundaries between humans and *tshomen* more readily dissolve. Aum Pema explains:

> In our dreams, the *tshomen* appears as a woman wearing a beautifully embroidered *kira* [Bhutan's female national dress] and combing her hair. To us locals, we see her in the shape of a person we already know. Still, we know that it is not that person, but the *tshomen*. An outsider [non-highlander] will see the *tshomen* as a person who is not known to him or her.

This differential perception of *tshomen* between highlanders and outsiders is also reflected in the rewards and reprisals the deities bestow, which predominantly affect herders. Dream encounters with *tshomen* readily impel social and ritual interaction in waketime. For instance, a gift from a *tshomen* during a dream signals an imminent *ngodrup*, whereas a *tshomen* who takes something from the dreamer speaks of impending loss. As the dream is invariably shared the next morning, both relieving the dreamer and turning the dream into a social fact (see Heneise 2016), it is responded to by climing the *tshomen*-lake ritually, through libation, or incense and milk offering, and chants on the shore. For instance, the dreamer might chant: "*Dhari chu dhi nang gi neydhag ga en ru, nga gi om tsang tokto dhang sang phuel dho. Nga lu jabchor chi na gi ra zel goh* [Today I offer you milk and incense. You are the only one who can protect me]."

Blessings, sometimes pre-announced in dreamtime, come in various forms. In addition to *tshoyak*, herders have received gemstones, particularly the precious turquoise. Yeshey reports:

> There is one female herder. She resides near the lake and daily offers milk to the *tshomen*. One day, the lake swirled and bubbled. As fresh water was streaming into the lake from uphill, she just ignored it. Meanwhile, she guided her yaks to the top of the hill to graze on fresh grass. When she returned sometime later to check on the swirling water, she found turquoise stones waiting for her. These are priceless and they were a clear gift from the *tshomen*, as the stones were not there earlier.

Tshomjay is another *ngodrup*. It refers to an experienced reality in which temporalities collapse in ways that are illustrative of the *tshomen*'s immorality, enabling the human recipient to see his or her past, or perhaps the future of this or subsequent lifetimes. This nonlinear experience of time is accounted for in Buddhist cosmology and bequeathed only to those who are unhindered by *drib*, usually while they are meditating near the lake. Herders report that a *tshomjay* experience is usually preceded by the water in the lake churning and turning milky. Yeshey's grandfather and aunt both experienced visions of their past and future, with the latter's *tshomjay* dominated by sheep. Thereafter, the size of her flock increased significantly. When she died, so did her sheep, almost instantly. Such fables still circulate widely in the highlands, although the herders say the experiences themselves are becoming increasingly rare "in this generation."

In contrast to these reports of benevolence, herders are also wary of aggressive, aggrieved *tshomen* who might deprive them of wealth and health. Such behavior is discussed in terms of *jigdra*, which translates as fear, horror, terror, fright, nightmare, and awesome power. This hierarchical, contextually disdainful disposition—which is common to both humans and deities—is always cast down from a position of social superiority. For instance, humans exhibit *jigdra* when explicitly expressing their superior status or seniority over subordinates; and a *tshomen* does the same when displaying displeasure at human presence by repeatedly causing the lake to churn, creating sudden waves, or summoning a thick fog.

In the highlands, the Jimilang-tsho and Lang-tsho lakes are visualized as father and son. Lang-tsho, the smaller of the two, is the son, fiercer and more dangerous than his more composed father. For many centuries, this lake was isolated, far from the trails of humans and yaks, and therefore remained unaccustomed to the presence of humans and the *drib* they carry. Now, though, Lang-tsho is within human purview and reacts strongly whenever a herder approaches. Sonam reports: "Time and again the lake starts vibrating and shivering. The leaves that float on it begin to move in an outward manner. Then a huge amount of water jumps out from the middle, just like a fountain." This is evidence of *jigdra*. It is the *tshomen* demonstrating her power, acting proudly toward humans, and signaling her reluctance to honor the *tshomen*–human social contract.

Jigdra can also be violent, or even fatal, as Sonam explains: "The area around the lake suddenly becomes so foggy that people lose their sense of direction. Then the *tshomen* lures them into the water. Only when the person has drowned will the lake leave the body at its banks." Alternatively, a *tshomen*-lake might chase humans by sending water beyond her banks through flooding. Hence, Sonam's parents advised him to run to higher ground whenever a *tshomen*-lake was about to overflow. Another herder, Pema, recalls:

> Back in the day, we had to pay *latsi* [the scrotum of a bharal or Himalayan blue sheep] and birds' feathers in tax to the government. For this reason, we had a *sharob* [community hunter]. One day, when the hunter was near a lake, he saw arrows made of thin bamboo growing all around it. Intrigued, he took out his knife and harvested them. Then he heard "*sheraab, sheraab*" coming out of the lake. He initially thought it must be an animal, and, being a hunter, he was not afraid. But when he looked again at the lake, he noticed that it was preparing to chase him, so he began running down the hill. When he finally rested, he checked his bag, only to find that the *tshomen* had retrieved her arrows.

But if *tshomen*-lakes can display *jigdra* by chasing people, herders can do likewise by chasing away lakes. "In my parents' time," Dorji recalls, "there was a lake that always caused disharmony and death. The *tshomen* was always harming our people. So it was decided that she had to leave." This was achieved by hanging the body of dead horse from a tree on the shore. So much *drib* seeped from the carcass that the *tshomen* could not bear it and fled, causing the lake to dry up.

Clime Change, Migrating Waters, and the Inversion of *Jigdra*

If we apprehend the dynamism of glacial lakes beyond a limited human time frame, they cease to be permanent bodies of water. Instead, they appear and disappear along with shifts in the cryosphere. Indeed, when we broaden our historical lens to encompass geological and paleo-climatic temporalities, we see countless lakes emerging and dying over time. This inherent shifting of bodies of water is inflected in the cosmological character of *tshomen*, who are known to occasionally evacuate their liquid palaces and take up residence elsewhere, taking the lake-water with them. Such evacuations are evident in significant drops in water level, or in a lake communicating the sudden absence of its *tshomen* by darkening in color. While lake migration speaks of the climatic agency of the earth, *tshomen*-lakes are now thought to be willfully disassociating from humans by fleeing anthropogenic transformation. For instance, Karma Ura (2001, 9) reported that the Buli-tsho of Bumthang relocated to Zhemgang in response to human-induced contamination. "*Tsho po-bay yar-so-nu* [The lake has shifted/migrated)," herders say when a lake disappears or overflows its banks.

Anthropogenic climate change is now rearranging the hydrological cycle in the Bhutan highlands, and across the Himalayas, by altering the balance and interactions between water in its solid, liquid, and vaporous states so that earlier patterns of glacial and snow melt, recharge, precipitation, evaporation, and temperature are no longer reproduced in a stable manner. When micro-climatic and ecological variables are plugged in, the drying up and overflowing of lakes (herders observe both) are integral to the anthropogenic highland hydrological cycle. Herders read and relay these changing eco-climatic conditions as a breakdown of human–*tshomen* relationality. When they report fewer *tshomen* sightings and blessings "in this generation," they imply that the current generation is failing to cultivate reverential and caring relations with *tshomen*-lakes.

This emergent imbalance and disconnect is part of a more widely experienced dissonance between various eco-materialities, as enlivened by deities, and contemporary humans in the Bhutan highlands (see Choki 2021; Dema 2021; Chapter 10, this volume). Pointing out a particular *tsho*, Ugyen recalls that it used to shrink to half its size during the winter. Now, though, it shrinks to the size of a puddle. Yaks walk on the lake bed, leaving dung that defiles the water when it returns in the summer. For Ugyen, this is a sign of the imminent departure of the resident *tshomen*. A well-known local *tship* (astrologer, divine messenger) confirmed this interpretation by divining that "water is fleeing the lake" because younger highlanders are abandoning the pastoral lifestyle and its emplaced cosmovision. Such changes interrupt the norm-worlds and rituals that previously bound *tshomen*-lakes to herders, and leave Ugyen's generation anxious about the future. He laments: "Nowadays, rains arrive suddenly and are very forceful, but they leave equally suddenly and don't return for a long time." As a result, groundwater recharge is disabled because of the slow recharge rate of aquifers, causing the majority of water to run off rather than linger to recharge the lakes from below. When nearby glaciers and snowlines have significantly retreated, or vanished altogether, the reduction in rainwater is exacerbated by the non- or delayed arrival of glacial and snow melt. Moreover, torrential downpours lead to excessive soil erosion and sediment deposits in highland lakes, forcing away much of the water and darkening any that remains. Ugyen links these drying and darkening waters to lifestyle changes among the younger generations, who are increasingly attuned to modern aspirations. Older herders, such as

Ugyen, identify new material desires, new buildings, and new development activities as new causes of *drib*. Meanwhile, the younger generations have little enthusiasm for community rituals and festivals, so there are fewer opportunities for individual and collective *drib* reduction through the reactivation and strengthening of human–deity relations, including with *tshomen*.

In terms of development and (climate) change, herders identify the arrival of concrete as the most characteristic/charismatic change in recent highland life. Compared to Bhutan's valleys, concrete still has a minimal presence in the highlands, but it is becoming ever more visible. Some herders propose a linear correlation between its arrival in the region and palpable changes in the local climate. In broad, pan-Himalayan terms, changes and stresses to water tables are augmented by the contemporary contagion of concrete, which is the most invasive "species" of the Himalayas (as it is globally), given that it is now the most widely used substance on earth, after water. "If the cement industry were a country," writes Jonathan Watts (2019), "it would be the third largest carbon dioxide emitter in the world with up to 2.8bn tonnes, surpassed only China and the US." As the foundation of modern life, concrete is not generally viewed as a problem, as it is not derived from fossil fuels; nor, in marked contrast to plastic, is it routinely found in human stomachs and blood, or tangled in trees or oceans. However, its production is the largest contributor to global carbon-dioxide emissions after coal, oil, and gas. What is more, and of greater relevance to this discussion, concrete is turning the land from green and blue to gray. And this graying of the landscape cascades into micro-climatic changes.

Reports from Sikkim (McDuie-Ra and Chettri 2020) and Tibet (Grant 2022) show the foundation of modern life in high-altitude lands is enacted into concrete, with various governmental, developmental, cultural, and aesthetic effects. This process is beginning to expand into patches of the Bhutan highlands, and herders are already discussing it with concern. Scholarly and policy accounts rarely discuss concrete's impact on the weather and climate. Yet, it is the abundance of this material—in buildings, pavement, and other surfaces—that makes cities warmer than surrounding rural areas. Cities also have a lower relative humidity because rainwater is not absorbed into the ground to be subsequently released into the air by evaporation, and transpiration is lower because cities contain so little vegetation. This contributes to cities' distinctive anthropogenic climate. Affiliated micro-level climatic changes are expected to result from the ongoing process of concreting the Himalayas. The graying of the landscape—which, at present, affects only relatively small parts of the Bhutan highlands—produces heat effects that interfere with micro-precipitation patterns and creates a barrier between the soil and rainfall, preventing water absorption. In this way, concrete suffocates the soil by covering it, reducing evaporation, which in turn leads to reduced precipitation.

Herders already foresee the consequences of the concreting of their surroundings. For instance, they explicitly link climate change to the rapid expansion of the built environment. To them, these developments are indicative of a broader shift—especially among younger generations—to materialist mindsets and aspirations at the expense of upholding shared, more-than-human norm-worlds. They cite the arrival of new roads, new constructions, and the transformation of trees into electric poles as prime causes of changing climatic and ecological conditions, and link these, in a near-linear manner, to the resentment of *tshomen* and other numinous resident beings. Such concerns caused one herder to object vehemently to any suggestion of road-building in the highlands:

> We cannot have roads come to our highlands because of the many *tshomen*-lakes. Roads will harm them. Roads will also bring more people who will cause disturbances and pollution to the lakes. We must leave our lakes in peace, as we have always done. Our lakes provide us with blessings. We cannot take the risk of losing our blessings to roads.

But if some highland lakes are drying up and darkening, others may soon overflow their banks as rising temperatures accelerate glacial melt and water inflow. For instance, Figures 3.4 and 3.5 depict the unprecedented growth of Tshomphu Lake between 2006 and 2016 (the period for which we were able to obtain reliable data).

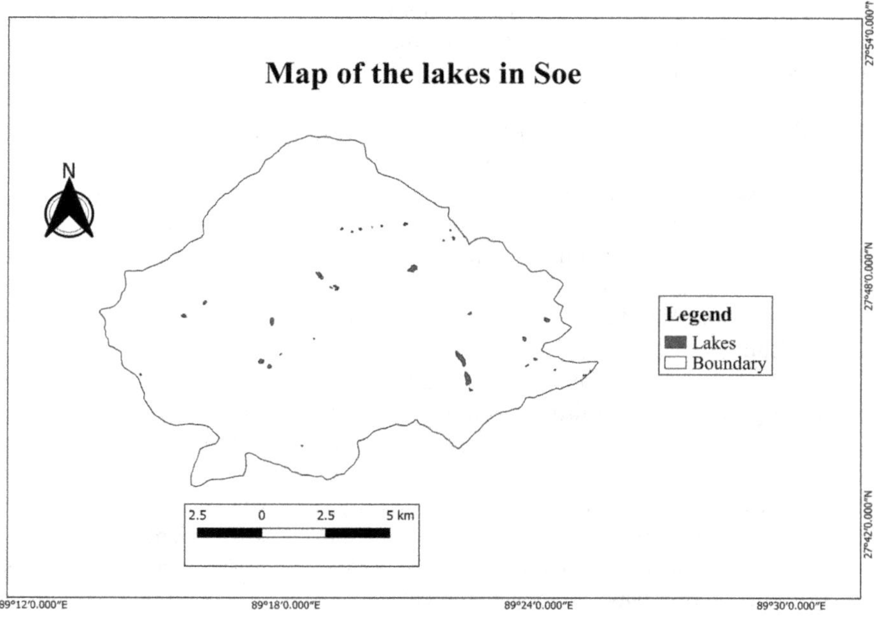

Figure 3.4 Map of lakes in Soe, Thimphu, Bhutan, 2006
Note: The largest lake on this map, located in the southeast, is Tshomphu Lake, which is mapped in Figure 3.5.
Source: The authors

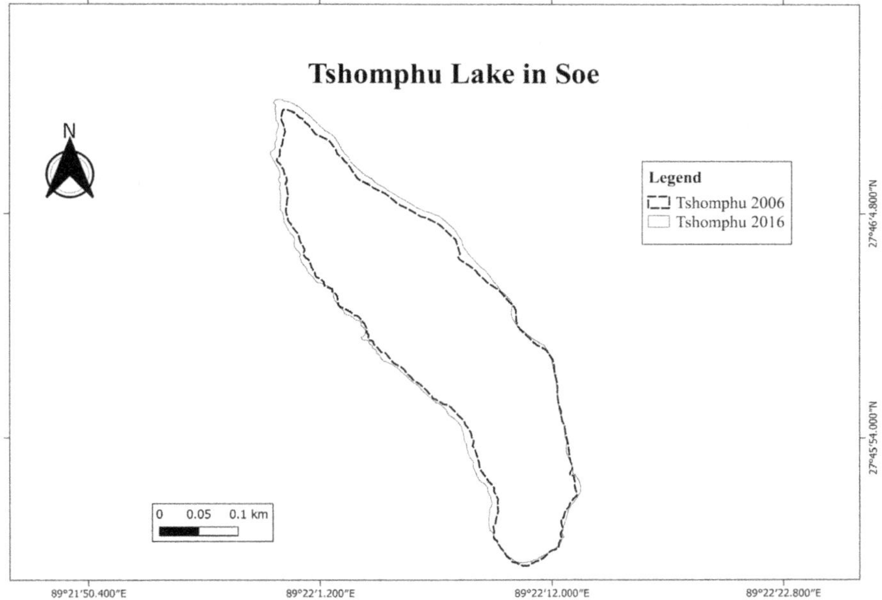

Figure 3.5 Map of the increasing size of Tshomphu Lake
Note: The black line shows the size of the lake in 2006; the gray line shows its increased size in 2016.
Source: The authors

These palpable changes in water, whether in the expansion or contraction of lakes, are read by herders as indicative of the changing interactive attitude between humans and *tshomen*. While highlanders are taught from a young age how to comport themselves near *tshomen*-lakes so that *tshomen* might favor them and the herding community at large, today a certain ambiguity has crept into the shared sensibilities and norms that were negotiated long ago by Padmasambhava.

In extreme cases, rather than appeasing the *tshomen*-lakes, humans may deliberately challenge the cosmological order by willfully displaying *jigdra* toward them in the form of verbal and non-verbal insults. A few years ago, the Thimphu Valley suffered a sudden shortage of water due to reduced rainfall. The harvest was threatened and drinking water became scarce. Knowing that *tshomen*-lakes are weather-makers, several desperate farmers trekked to a large lake in the northwest of the valley and polluted it in a bid to provoke the *tshomen* into producing rain. The rains duly arrived, but they were not the mild downpours the farmers had anticipated. Indeed, the deluge was so intense that it threatened to flood not only the fields but the city. Moreover, the farmers fell gravely ill and were soon admitted to hospital. As their health continued to deteriorate, their families approached the local monks and told them what their menfolk had done. The monks immediately traveled up to the lake, where they appeased the *tshomen* with offerings and chants. The rain ceased and the farmers recovered.

Highland herders are also sometimes driven to express *jigdra* toward lakes, especially during droughts. For instance, Nidup confessed that he and others throw horses' bones into an authoritative *tshomen*-lake in the hope that the *tshomen* will demonstrate her anger by producing much-needed rain. When enough rain has fallen, they remove the bones from the lake and ritually apologize to the *tshomen* with offerings. Such behavior is unprecedented and risks jeopardizing the more-than-human social contract that facilitated human settlement in the highlands in the first place. Yet, it reflects the desperation of herders in the highlands and farmers in the valley, all of whom are grappling with a new, anthropogenically affected hydrological cycle.

Notes

1 This chapter has grown out of our multiple-year engagement with herders, yak, mountains, lakes, and other eco-materialities in the Bhutan highlands. Ethnography was conducted in the highlands in several phases—some short (days or weeks), others longer (two to three months)—of collaborative fieldwork that began in 2018. This was complemented by interviews with herders who have settled in the Thimphu and Paro valleys and others who descend from higher altitudes to spend the cold winters there. We express our gratitude to the research assistants who helped with this research: Pema Choden, Sonam Tenzin, Kinley Wangmo, Kinley Choki, Tshering Pelzom, Pema Gyeltshen, and Kinley Dorji.

2 *Tshomen* is colloquially translated as mermaid in Bhutan. However, there are other translations. Etymologically, *tshomen* (or *tshomem*) refers to a lake spirit or lake goddess; *tsho* means lake or ocean and *mem* means medicine, remedy, or herb. Karma Ura (2001) refers to *Tsho mammo*, with *mammo* translating as woman, while Thinley Dema (2021) has presented *tshomem* as lake deities. In our fieldwork, the translation of *tshomen* as mermaid was found to be the most common. The pronounciation of *tshomen* also varies somewhat across local dialects.

References

Bajracharya, Samjwal Ratna, Sudan Bikash Maharjan, and Finu Shrestha. 2014. "The Status and Decadal Change of Glaciers in Bhutan from the 1980s to 2010 Based on Satellite Data." *Annals of Glaciology* 55(66): 159–166.

Bennett, Jane. 2010. *Vibrant Matter: A Political Ecology of Things*. Durham, NC: Duke University Press.

Choki, Kinley. 2021. "Cordyceps, Climate Change and Cosmological Imbalance in the Bhutan Highlands." In *Environmental Humanities in the New Himalayas*, edited by Dan Smyer Yü and Erik de Maaker, 152–166. Abingdon: Routledge.

Cryosphere Services Division, National Center for Hydrology and Meteorology. 2021. *Bhutan Glacial Lake Inventory 2021*. www.nchm.gov.bt/attachment/ckfinder/userfiles/files/Bhutan%20Glacial%20Lake%20Inventory%202021.pdf.

Dema, Thinley. 2021. "Eco-Spiritual and Economic Perspectives in Bhutan's Haa District." In *Environmental Humanities in the New Himalayas*, edited by Dan Smyer Yü and Erik de Maaker, 66–80. Abingdon: Routledge.

de Wolff, Kim, Rina C. Faletti, and Ignacio López-Calvo. 2021. *Hydrohumanities: Water Discourse and Environmental Futures.* Oakland: University of California Press.

Grant, Andrew. 2022. *The Concrete Plateau: Urban Tibetans and the Chinese Civilizing Machine.* Ithaca: Cornell University Press.

Haberman, David L. 2021. *Understanding Climate Change through Religious Lifeworlds.* Bloomington: Indiana University Press.

Heneise, Michael Timothy. 2016. *Life and Landscape of Dreams Personhood, Reversibility and Resistance among the Nagas in Northeast India.* https://era.ed.ac.uk/handle/1842/25523.

Huber, Toni. 1999. *The Cult of Pure Crystal Mountain: Popular Pilgrimage and Visionary Landscape in Southeast Tibet.* New York: Oxford University Press.

Huber, Toni and Poul Pedersen. 1997. "Meteorological Knowledge and Environmental Ideas in Traditional and Modern Societies: The Case of Tibet." *Journal of the Royal Anthropological Institute* 3(3): 577–597.

Irvine, Richard D.G. 2020. *An Anthropology of Deep Time: Geological Temporality and Social Life.* Cambridge: Cambridge University Press.

Kumar, Mithun, Ayad M. Fadhil Al-Quraishi, and Ismail Mondal. 2021. "Glacier Changes Monitoring in Bhutan High Himalaya Using Remote Sensing Technology." *Environmental Engineering Research* 26(1): 190255.

Lahiri, Dutt. 2015. "Beyond the Water–Land Binary in Geography: Water/Lands of Bengal, Revisioning Hybridity." *ACME: An International Journal for Critical Geographies* 13(3): 505–529.

Latour, Bruno. 1991. *We Have Never Been Modern.* Cambridge, MA: Harvard University Press.

Linton, Jamie. 2010. *What Is Water? The History of a Modern Abstraction.* Chicago: Chicago University Press.

McDuie-Ra, Duncan and Mona Chettri. 2020. "Concreting the Frontier: Modernity and its Entanglements in Sikkim, India." *Political Geography* 76: 102089.

McPhee, John. 1999. *Annals of the Former Worlds.* New York: Farrar, Strauss, and Giroux.

Mentz, Steve. 2023. *An Introduction to the Blue Humanities.* London: Routledge.

Paerregaard, Karsten. 2023. "Climing the Andes: Vertical Complementarity, Transhuman Reciprocity, and Climate Change in the Peruvian Highlands." In *Storying Multipolar Climes of the Himalaya, Andes, and Arctic: Anthropogenic Climate and Shapeshifting Watery Worlds*, edited by Dan Smyer Yü and Jelle J.P. Wouters, 52–68. Abingdon and New York: Routledge.

Pandit, Maharaj K. 2017. *Life in the Himalaya: An Ecosystem at Risk.* Cambridge, MA, and London: Harvard University Press.

Phuntsho, Karma. 2013. *The History of Bhutan.* Delhi: Penguin Random House India.

Pommaret, Francoise. 2014. "Bon in Bhutan: What is in the Name?" In *Bhutanese Buddhism and its Culture*, edited by Seiji Kumagai, 113–126. Kathmandu: Vajra Publications.

Saikia, Arupjoyti. 2019. *The Unquiet River: A Bibliography of the Brahmaputra.* Oxford: Oxford University Press.

Scott, Robert W. and Floyd A. Huff. 1997. *Lake Effects on Climatic Conditions in the Great Lakes Basin.* Champaign: Illinois State Water Survey.

Smyer Yü, Dan. 2020. "The Critical Zone as a Planetary Animist Sphere: Etho-graphing an Affective Consciousness of the Earth." *Journal of the Study of Religion, Nature, and Culture* 14(2): 271–290.

Smyer Yü, Dan. 2021. "Situating Environmental Humanities in the New Himalayas: An Introduction." In *Environmental Humanities in the New Himalayas Symbiotic Indigeneity, Commoning, Sustainability*, edited by Dan Smyer Yü and Erik de Maaker, 1–24. Abingdon: Routledge.

Smyer Yü, Dan. 2023. "Multipolar Clime Studies of the Anthropocenic Himalaya, Andes and Arctic." In *Storying Multipolar Climes of the Himalaya, Andes and Arctic: Anthropogenic Climate and Shapeshifting Watery Worlds*, edited by Dan Smyer Yü and Jelle J.P. Wouters, 1–26. Abingdon and New York: Routledge.

Sörlin, Sverker. 2015. "Cryo-History: Narratives of Ice and the Emerging Arctic Humanities." In *The New Arctic*, edited by Birgitta Evengård, Joan Nymand Larsen, and Øyvind Paasche, 327–339. New York: Springer.

Sousounis, P.J. 2001. "Lake Effect Storms." *Encyclopedia of Atmospheric Sciences* 1104: 1115.

Su, Dongsheng, Lijuan Wen, Xiaoqing Gao, Matti Leppäranta, Xingyu Song, Qianqian Shi, and Georgiy Kirillin. 2020. "Effects of the Largest Lake of the Tibetan Plateau on the Regional Climate." *Journal of Geophysical Research: Atmospheres* 125(22). https://agupubs.onlinelibrary.wiley.com/doi/full/10.1029/2020JD033396.

Tashi, Kelzang. 2023. *World of Worldly Gods: The Persistence and Transformation of Shamanic Bon in Buddhist Bhutan*. Oxford: Oxford University Press.

The Third Pole. 2021. "Are Bhutan's Monsoon Landslides Becoming More Deadly?" July 14. www.thethirdpole.net/en/climate/bhutan-landslides-climate-change/.

Tshewang, Ugyen, Michael Charles Tobias, and Jane Gray Morrison. 2021. *Bhutan: Conservation and Environmental Protection in the Himalayas*. Cham: Springer.

Ura, Karma. 2001. "Deities and Environment: A Four-Part Series." *Kuensel: Bhutan's National Newspaper*, November 26.

Wangchuk, Sonam, Tobias Bolch, and Benjamin Aubrey Robson. 2022. "Monitoring Glacial Lake Outburst Flood Susceptibility Using Sentinel-1 SAR Data, Google Earth Engine, and Persistent Scatterer Interferometry." *Remote Sensing of Environment* 271: 112910.

Watts, Jonathan. 2019. "Concrete: The Most Destructive Material on Earth." *Guardian*, February 15. www.theguardian.com/cities/2019/feb/25/concrete-the-most-destructive-material-on-earth.

Wouters, Jelle J.P. 2021. "Relatedness, Trans-Species Knots and Yak Personhood in the Bhutan Highlands." In *Environmental Humanities in the New Himalayas*, edited by Dan Smyer Yü and Erik de Maaker, 27–42. London: Routledge.

Wouters, Jelle J.P. 2023. "Where Is the 'Geo'-Political? More-than-Human Politics, Polities, and Poetics in the Bhutan Highlands." In *Capital and Ecology: Developmentalism, Subjectivity, and Alternative Life-Worlds*, edited by Rakhee Bhattacharya and Amarjit G. Sharma, 181–202. Abingdon and New York: Routledge.

4 Storied Toponyms in Bhutan

Affective Landscapes, Spiritual Encounters, and Clime Change

Kinley Dorji

As a student of environmental science, I learned to understand the environment through the scientific and technical study of the earth's physical, chemical, biological, and geological systems and processes. I was taught to ask questions about nature and climate with the methodological approaches and tools of the sciences, as well as to speak in a language of taxonomies, numerical models, and abstract discourses. In this process, I acquired an elaborate technical jargon to represent the environment and to account for changes therein. I accepted this scientific knowledge as part of my educational advancement and as preparation to become an expert on environment and climate (change) in my native Bhutan.

While the scientific knowledge I thus acquired continues to aid my understanding of climatological and ecological history, patterns, and change, when I started teaching and researching human–environment relations back in Bhutan I gradually realized the shortcomings and limitations of science-only perspectives. To my students, the environment was sentient and storied, and their thinking reflected this knowledge. Their backgrounds were similar to my own in rural eastern Bhutan. However, the syllabi, assignments, and examinations we bestowed upon them did not allow for our emplaced and embodied knowledge to be considered seriously. Instead, the focus was on changing their indigenous knowing, relating, and seeing to a scientific assessment of the environment and climate. This made me more acutely aware of a wedge between the environments in which my students and I lived and the environment I was obliged to teach.

The longer I taught, the more torn I became between the affective, indigenous knowledge my students and I share and the theories and models prescribed in our curriculum. My growing resentment was not that scientific knowledge was inaccurate—far from it—but that science-only perspectives did not do justice to the lived experience and the understanding that Bhutanese acquire through their everyday involvement with the landscape. My technical–scientific knowledge, tools, and language increasingly began to appear as distanced and detached from the world I shared with my students and other Bhutanese. It was during this time that I first encountered Robin Wall Kimmerer's (2013) monumental *Braiding Sweetgrass: Indigenous Wisdom, Scientific Knowledge, and the Teachings of Plants.* What struck me

DOI: 10.4324/9781003484394-4

was the ingenious ways in which Kimmerer, an indigenous scholar herself, was able to weave scientific knowledge with ancient indigenous wisdom through the intertwining of "science, spirit, and story" (Kimmerer 2013). Her drawing of lines and connecting of dots between scientific and indigenous knowledge suggests that the two are not oppositional, but can in fact complement one another to create more inclusive and capacious understandings of the environment and climate (change) (see also Wouters 2023a).

Like Kimmerer's, the environment in which I grew up was full of stories. Whether a mountaintop, valley, river, cliff, or cave, there always seemed to be a story attached to them. These stories were distinctly multispecies in their orientation and included, besides humans, deities, spiritual masters and sages, animals, and trees. These stories were also not just descriptions; they included moral and spiritual lessons, as well as prescriptions and prohibitions about social behavior. They were deeply grounded in our Bon–Buddhist heritage and signaled the intimate relations between the physicality of the environment and spiritual and religious lifeworlds. It was my return to these stories of my upbringing that made me realize that, while scientific–technical knowledge offers a privileged window into the environment and climate, science-only knowledge was not doing sufficient justice to the diverse, lively, and affective experiences and knowledge traditions that exist in Bhutan.

My turn to clime studies is grounded in this experienced need to combine scientific understandings with lively ethnography and multispecies accounts of actually lived climate change. Here, I take inspiration from Shepherd and Truong (2023), who seek to complement the natural sciences with the social sciences and the humanities by combining the storylining approach of climate science with the storying approach of the environmental humanities (see also Smyer Yü 2023, 21). Similarly, the natural scientist Vandana Singh (2023) shows how an inter-scientific spirit can be productive in both knowing and teaching climate change. She transcended her earlier, physics-only understanding and teaching of climate change by integrating the perspectives of the indigenous Inupiaq in the Arctic, as well as by incorporating ethnographic and anthropological perspectives. Her classrooms on climate change are now distinctly transdisciplinary by merging scientific–technological knowledge with the epistemological and psychosocial action dimensions of climate change. In her own evaluation, this led to more engaged and affective ways of teaching and learning about climate change in her classroom.

Singh teaches in the United States, and her students are predominantly American. To them, the exercise was one of adopting radical openness in understanding climate change by expanding their horizons of what counts as knowledge. In contrast, my task in Bhutan is not to introduce indigenous knowledge, but to encourage my students to bring their already embodied indigenous knowledge into climate (change) discussions—that is, to allow the sciences and indigenous knowledge to fertilize each other. It is in this spirit that I adopt a storying-climes approach in this chapter to complement scientific knowledge with richly layered, affective, experiential, and multispecies approaches to understanding environment and climate in Bhutan.

Introduction: Storied Toponyms in Bhutan

In Bhutan, stories multiply as one clim(b)es the landscape. They intersect and interact with one another in a dense network of intertwined histories, spiritual encounters, and cultural memories. If a clime can be understood as the mutually embodied relations of climate, place, nature, culture, humans, and other-than-humans, stories in Bhutan are constitutive of climes because they co-create the places and worlds in which Bhutanese humans and other-than-humans participate. Stories are therefore an intimate part of the worlds they describe; they frame ethics, beliefs, understandings, and relationships between humans, other-than-humans, and the materiality of place. I offer three vignettes to illustrate how stories and places—or climes—are co-constituted in Bhutan.

Fieldwork for this chapter included the Bhutan highlands, which borders the Tibetan Autonomous Region. Until the formal closure of this border in 1959, following the Chinese annexation of Tibet, these highlands, despite their harsh terrain, were part of a lively trade corridor between Tibet, Bhutan, and the Assam plains below. It was through this corridor that Zhabdrung Rinpoche Ngawang Namgyal, recognized as the unifier of Bhutan, arrived in 1616. Zhabdrung Rinpoche is an honorific reference to Ngawang Namgyal and translates as "at the feet of a rare and precious personality a revered bow can be placed." Zhabdrung was a reincarnation of the Buddhist master Künkhyen Pema Karpo who hailed from Kongpo in Tibet. Following a monastic feud that included a threat on his life, Zhabdrung fled Tibet, heeding a call from his protective deities who showed him the way to Bhutan. The local Bhutanese understood the highness and rarity of such a personality in their land and submitted themselves to him.

Zhabdrung's arrival in Bhutan and his spiritual and political legacy have been recounted in detail by Karma Phuntsho (2013, 229). Here, I emphasize how his presence became emplaced in the landscape through storied toponyms. A yak-herder from the village of Lingzhi explains:

> Zhabdrung arrived via Laya. He first entered the valley of Chheypuelsa [commonly pronounced Cheybesa]. The herders offered Zhabdrung the tongue of a yak as a special meal. Now, *chey* means tongue and *phuel* means offering. Zhabdrung continued his journey and arrived in what is today Gonyuel Village. *Gong* translates as "eggs," and this was the offering the herders made to Zhabdrung there. The next hamlet Zhabdrung visited is now known as Shayuel. *Sha* means "meat" and *yuel* means village. It was in this village that Zhabdrung was given yak meat to eat.[1]

Zhabdrung's journey thus became emplaced in the landscape through toponyms that are still used today, and which, in this case, function as markers of a historical event.

Zhabdrung is not the only saint whose arrival in Bhutan became emplaced in the landscape through toponyms. My second vignette concerns the arrival of a compassionate Indian prince named Drime Kunden who was banished to Durihashang (Black Mountain Range in central Bhutan) by his father, the king, for giving *Nob Gaduen Pungjom* (a wish-fulfilling jewel) to an enemy soldier disguised as a beggar. On his way into exile, his compassion for the poor prompted him to give away all his possessions, including his children, his wife, and even his own eyes to, respectively, an old childless couple, a businessman, and an old blind man. These acts of kindness are remembered today in toponyms. For instance, in Trongsa, central Bhutan, there is a place named Bubja, which was earlier called Bujin, to signify the place where Drime Kunden gave away his children—*bu* means "children" and *jin* means "to give." Similarly, the ancient trail that leads to the Black Mountains passes a place named Chenray. However, before this pronunciation was adopted, it was called Chenjin to commemorate that this was where the prince donated his eyes to a beggar—*chen* is an honorific reference to "eyes." Thus, local place names became interwoven with Drime Kunden's actions.

The third illustration relates to Padmasambhava, also known by his devotees as Guru Rinpoche or the Second Buddha. This historical and spiritual figure is central to the epic of modern Bhutan because he secured the land for Buddhism by taming earlier human-averse deities under a "sacred oath," turning them into protectors of Buddhism and initiating more-than-human "norm worlds" (Banerjee and Wouters 2022, 110) between humans and terrestrial deities. In the autumn of 2008, I visited Aja Ney, a place in eastern Bhutan that Padmasambhava visited. It has celestial abodes emplaced into the mountains, indicating the spiritualization of geology. Visualized and immortalized in the rocks are the hat, back, hand, foot, and scepter of Padmasambhava. Aja Ney is deeply immersed in stories of Padmasambhava's accomplishments, including how he tamed the local malicious spirits. To reach it, my father, several of his friends, and I had to walk for two days. We were pilgrims, devoutly ready to walk in Padmasambhava's footsteps. We spent four days listening to our local guide's stories about him. He explained that the place is called Aja Ney because of the 100 self-arisen *a* (ཨ) syllables on the rock face on the bank of the Aja River, opposite Padmasambhava's meditation cave. According to the guide, these signify Padmasambhava's meditative accomplishments. *Aja* translates as "one-hundred *a* syllables" and *ney* as "pilgrimage site." Today, Aja Ney is owned and protected by the local deity Lumo Takdongma, who is a subterranean Naga (serpent) from the waist down, with a human torso and the face of a tigress.

Aja Ney is just one place out of many in Bhutan where Padmasambhava's presence and accomplishments are engraved in stories and toponyms. For now, suffice it to note that the arrival of Zhabdrung, Drime Kunden, and

Padmasambhava have become emplaced in the Bhutan landscape through storied toponyms that today can be read as a history text. These toponyms exemplify the overall contention of this chapter—namely, that Bhutan's landscape is conceptually and perceptually enlivened and experienced through stories, many of which, in turn, are condensed into toponyms. Bhutanese often speak of the landscape by invoking these toponyms. My storying-climes approach affirms them as clime-makers: they are markers of history, meaning-making, and spirituality. I thus adopt an experiential approach to highlight the entanglements of history, place, and subjectivity. This allows for an affective, relational, and multispecies reading of the geomorphological texture of the Bhutanese landscape, making it appear as a living text. In this context, clim(b)ing becomes an exercise in seeing, sensing, and learning from the stories places embody.

Climes as Affective Texts

My storying-climes approach, in a sense, considers the landscape as a text that must be both understood and heeded to ensure fruitful co-dwelling between humans, other-than-humans, and the earth itself. Across Bhutan, deities, of both Bon and Buddhist origin and heritage, are the dominant nonhuman presence, and their behavior, rules, and interactions with humans are everywhere contingent on their location.

Approaches that seek to understand the landscape as a text so as to highlight it as a signifying system through which social, cultural, and spiritual systems are experienced and communicated have been critiqued for promoting disembodied knowledge. Offering a strictly materialist approach, the landscape anthropologists Tilley and Cameron-Daum (2017, 4–5) write: "It is through material experience that we can understand the ideological nature of these representations, the manner in which they quite literally frame the landscape, far better than by undertaking any desk-bound analysis." Tilley and Cameron-Daum (2017, 5) present a materialist approach to the landscape as a "return to the real":

> walking is not a text, cutting down a gorse bush is not a text, training to be a soldier is not a text, a body is not a text, hills and rivers and trees are not texts … The move is from representation to the materially grounded messiness of everyday life and the minutiae of material practices that constitute it.

Here, I argue for a middle ground that connects the two approaches. First, reading the texture of the land as a text is not a disembodied or "desk-bound" exercise in Bhutan. To understand the meaning and instruction of these "earth texts," one must be physically there so that one may see, smell, sense, and touch the materiality of the land. Here, it should be noted that I understand "text" in the broadest possible sense, including scriptures, stories (both orally transmitted and written down), and geo-ecological textures. To frame the landscape of Bhutan as text—that is, to practice a storying-climes approach—is to take seriously the ethnographic fact that places in Bhutan are emplaced precisely by stories. These texts, therefore, cannot be abstracted from the materiality of the landscape; rather, texture and text merge into one. This applies particularly to storied toponyms. Regarding the significance of toponyms, Karma Phuntsho (2013, 1–2) writes:

> The Bhutanese have etymologies for nearly all their place names. Throughout Bhutan, one can even today find a rich tradition of using toponyms to tell historical and religious narratives. Many stories of people's origin and local cultures are told by stringing together place names. When one listens to such a narration, the whole landscape comes back to life as a stage for legendary and historical events. Through the toponyms, we get glimpses of the elusive past and the names serve as mnemonic tools to remember local legends and histories.

In their everyday conversations, Bhutanese frequently talk in terms of the stories that places, in a profound sense, emplace. While the physicality of these places results from Deep Time geological and climatic processes, in affective, relational, and multispecies terms they come into being through richly layered and orally transmitted narratives that speak of their histories, spiritual guardians, and ethical qualities.

In a field-defining contribution to Native American studies, Keith Basso (1996) explores how the Western Apache envision and enact places. He shows how, among the Apache, notions of morality, memory, and history are not only about events or the actual storyline, but significantly focus on where in the landscape these events took place. Toponyms are created to emplace and enliven these stories, thus turning stories into place-makers. Among the Apache, the storied imagination is so pivotal that their sense of being, relating, and knowing is everywhere emplaced in landscape features. From this it follows that any meaningful understanding of Apache epistemological culture and cultural ontology must be approached through their practices of place-making. In day-to-day life, Apache storytellers invoke toponyms to instruct, guide, correct, and admonish other Apaches, and these toponyms are so well and widely known that the listeners immediately understand the message that is being relayed. Basso calls this "speaking with names" and explores the specific ways in which place-based narratives guide the Apache in how to act wisely.

As in Basso's work, toponyms in Bhutan, as enacted through stories, emplace in the landscape past historical, cosmic, and religious events from which moral instructions and spiritual lessons can be readily drawn. These stories thus shape conceptions not only of the landscape but of society and of living well with others—humans and other-than-humans alike. The latter include plants, animals, mountains, lakes, rivers, and caves, all of which are experienced, revered, and propitiated as spiritual entities. Storying Bhutan's landscape therefore starts with the affirmation that the physicality of the landscape and environmental flows is everywhere complemented by other-than-human presences and agents. If I ask my students to describe the environment of their natal villages, most of them invoke other-than-human beings to indicate and locate their village. This is avowed in the everyday use of the Bhutanese adjective *mi mayin*, which closely translates as "other-than humans," seen and perceived as sentient and intricately involved in the everyday lifeworld. Karma Ura (2001, 1) writes:

> Environment [in Bhutan], in its widest sense of the term, is the domain of occupants other than plants, animals and human beings. There are other beings to whom geosensitive areas of our country have been ascribed, and to whom, beyond human ownership, they belong.

Karma Ura calls these beings "immortal owners or landlords" and explicitly links their presence to environmental conservation. Humans, indeed, must cultivate amicable and respectful relations with these deities in order to settle, survive, and flourish in *their* places. Elizabeth Allison (2019) invokes this inner logic through her coinage of "deity citadels"—parts of the environment inhabited and owned by deities. She then draws relations between these deity citadels, development, and resource use:

> The presence of a deity citadel is sufficient in some locales to cause the diversion or reconsideration of human construction and resource use. By grounding spiritual beliefs in specific sites of the landscape, the citadels of deities sanctify the landscape, becoming nodes of resistance and resilience that support the Bhutanese in inhabiting their own internally-consistent cosmology.
>
> (Allison 2019, 268)

Affirming these insights, this chapter's storying-climes approach further enlivens the affective and more-than-human storied existence and co-constitutive relations between humans and other-than-humans in conversation with the materiality of Bhutan's landscape. My contribution to clime studies thus lies in bringing to the fore geo-ecological relationality to affirm how geological histories and Deep Time climatic sculpting antecedently provide the material context for multispecies liveability. In addition, my approach recognizes how complex forms of affective ecological relationality—in terms of organisms' practices, inventions, experiments, and ethics (including those of humans)—jointly craft multispecies lives and worlds. In this chapter, I understand "multispecies" in its broadest possible sense. While, in its physical and biotic sense, the term refers to a habitat of humans, animals, and plants, here it also includes mountains, glaciers, cliffs, caves, and rivers that are experienced and revered as the embodiments of powerful earth spirits throughout Bhutan.

Climes thus always involve multispecies entwinements, and patterns, enacted through subjectivity and experience, relate affective bonds and affordances of a living earth, and rotate around living histories and imagination. In this way, climes are at once geophysical, chemical, and ecological. But they are also—especially—relational and affective as they attune to the ethos, consciousness, and soul of the landscape, as I seek to capture here through the analytic of storied toponyms. In my approach, I highlight the "storied" and "multi-sensory" imagination that affirms Bhutan's distinctly affective "geo-ecology," or the interplay between geological forces, ecological relations, and affective human and other-than-human inhabitations. The case-studies and narratives below substantiate how, in Bhutan, climes enact as stories, while stories enact as climes.

Padmasambhava and Affective Geomorphology

Geomorphology invokes a way of knowing through the study of landforms and landform evolution. In qualitative terms, it refers to the description of landforms. Quantitatively, the focus is process-based and describes forces acting on terrestrial earth surfaces to produce landforms and landform change. Here, I add the word "texture," making it geomorphological texture, to supplement this scientific approach with affective, storied, and multi-sensory ways of knowing the landscape. Texture adds to the physical materiality of the earth the notion of "feel," which is integral to the "multi-sensory imagination," by which I mean ways of sensing and storying the landscape beyond the strictly material and through sensorial engagement. Chao and Enari (2021, 39) write:

> the multi-sensory imagination takes as its starting point the literal grounds in and upon which planetary life arises, transforms, senses, and senesces … [It is an approach that] moves away from framing the imagination as a purely conceptual or ideological praxis, conjured by our (human) minds and detached from our (more-than-human) surroundings.

Bhutanese, indeed, continuously "sense" the landscape, and their sensing at once refers to bodily affects and is informed by animated stories.

Thus, whereas knowing through geomorphology privileges a distant, disembodied, and overall abstract understanding of the material landscape, the multi-sensory imagination is an epistemologically specific immersion in the texture, feel, touch, sound, and smell of the physical earth. The physical texture of the Bhutan landscape antecedently provides the material context of multispecies life in the sense that it enables and sustains specific multispecies communities (including those of humans) in specific geo-ecological gradients. But more than this, it also sets the stage for how humans clime the landscape by orienting their spiritual, social, and ritual life to it.

In Bhutan, the texture of the land is habitually approached and translated through "ritual geology." Robyn D'Avignon (2022) coined this phrase to apprehend the mutual constitution of geological knowledge, including the practices, prohibitions, and cosmological engagements of, in his ethnography, African *orpailleurs* (gold-miners who use traditional tools and knowledge) with the physical earth. These include

> a set of expressions or gestures or a code of communication; the attempt to demarcate, and to make activities consist with, preceding cultural practices; rules that are imposed on, and meant to restrict, human action and interaction; and shared symbolism that is sacred in nature.
>
> (D'Avignon 2022, 5)

I find the phrase "ritual geology" useful in the context of Bhutan, for it captures the affective and more-than-human relational narratives of a landscape in which geological constructions such as mountaintops, cliffs, and caves are experienced and engaged in ritual and cosmological terms.

The rituality and spirituality of geology are emplaced in the act of pilgrimage, which many Bhutanese practice with devotion. When I was about 12 years old, my parents took me on a pilgrimage to important temples (*lhakhang*) across the country. As in Garhwal (Chapter 8, this volume), pilgrimage in Bhutan is an act of spiritual clim(b)ing of particular geo-ecological places entwined with affective, cosmological, and spiritual significance. I distinctly recall our visits to Jampa Lhakhang in Bhumtang and Kyerchu Lhakhang in Paro. My father told me that these two temples were Bhutan's most sacred and important. "It is here that our history started," he said, explaining:

> These temples were built by the seventh-century Tibetan Emperor Songtsen Gampo. It is said that he had a Chinese wife known by the name Azhi Jaza who had a fine knowledge of *tsi* [astrology, geomancy, and divination]. It is said that when Azhi was on her way to Lhasa, the capital of Tibet, to marry the king, the cart that carried her got stuck in the mud. As the cart remained immovable, she was afflicted by the thought of inauspiciousness. To understand the cause of this inauspicious sign, she abruptly pulled out her *datho* [a geomantic mathematical chart] and started surveying the land. Her study revealed that Tibet was geographically situated on a supine demoness. Azhi found out that Lhasa was located exactly on the demoness's chest, with the heart at Othang Lake. After her study, Azhi Jaza is said to have informed the king that 108 temples must be constructed in a single day above the vital parts of the supine demoness, to pin her to the land and cast away all the misfortune that would otherwise be caused to the king, his country, and the people of Tibet. This gigantic supine demoness spread her body and limbs beyond Tibet, and her left leg fell on Bhutanese soil. Heeding the queen's request, the compassionate king built 108 temples in a day. Jampa

> Lhakhang and Kyerchu Lhakhang are two of them, built, respectively, on the demoness's left knee and foot.

Years later, as a student, I found my father's story in Karma Phuntsho's (2013) monumental book *The History of Bhutan*. In it, he describes Songtsen Gampo as "one of the greatest kings of ancient Tibet"; how his wife (who is named as "Wencheng" in his account) conducted a geomantic survey of the land; how this revealed the presence of a "demoness's body stretched across the Himalayan landscape causing it to breed savagery and diabolic forces"; and how, for Buddhism to flourish, this "demonic landscape had to be tamed through building a series of temples" (Phuntsho 2013, 79). Jampa Lhakhang and Kyerchu Lhakhang were part of this vast temple-building project. Now, visualizing and engaging the geomorphology of the land as a demoness tamed by temples was never part of my scientific training. Yet, it is the way most Bhutanese conceptualize and perceive the environment and their history and habitation within it. Also in this spirit, Wouters (2023b, 190) argues that geomancy readily supersedes geography in experiential, relational, and terrestrial terms in Bhutan: "across its gradients, random arrangements in the landscape, whether the shape of a water-body, pebbles piled up, or the form of a mountain peak, are anything but random, but form an animated script that must be understood." Such perspectives are likewise central to Bhutan histories written in Dzongkha, in which the landscape is variously affirmed as sentient, agential, and animate (Tshering 2023, 12).

Ap (Father) Dawa Nidup is a retired herdsman who for many decades co-dwelled and co-migrated with yaks and horses in the highlands of Soe. He sits on a small mound of soft earth, which he knows is the warmest spot in Ngabithang (a flat area with five stupas). He holds a rosary in his left hand, pulling the beads one after the other with his thumb while muttering a prayer. He has lived with the mountain Jomolhari (see Chapter 10, this volume) for the last seven decades. Pointing to the mountain that faces Jomolhari, he tells me it is the *naykhang* (citadel) of Ap Chungdu (a terrestrial deity), the first of a chain of mountains that run down to a place called Zombathang, where there is a uniquely shaped ridge that the locals revere as a representation of Guru Rinpoche's Zhingkham (heavenly abode). In Buddhist terminology, this place is known Zangdog Pelri. He then tells me that Guru Rinpoche visited the place and blessed it. "The sight of it will prevent us from being born in the three hell realms," he says, joining his palms together to make a lotus bud mudra and then, as a sign of deep reverence, lifting his hands to his forehead. I ask what he meant when referring to Guru Rinpoche's Zhingkham. He replies, "Soe is a beautiful place. It is blessed by many Buddhist masters. This place has the power to provide knowledge and awaken the primordial essence of enlightenment, which is the nature of Guru Rinpoche himself." His sense of the landscape, the way he sees and experiences it, is thus directly informed by Guru Rinpoche's visit, which took place well over a thousand years ago.

As noted above, Guru Rinpoche—often referred to as Padmasambhava in Bhutan —was a guru from Pakistan/India who introduced tantric Buddhism to Tibet and Bhutan. He reportedly visited Bhutan twice in the eighth century, traveling across the country, subduing malevolent spirits, meditating, hiding treasures, and instructing the locals in Buddhist doctrine. Padmasambhava's actions were so foundational to the making of modern Bhutan that they are everywhere emplaced in the landscape, and remembered and transmitted through stories. So ubiquitous are the toponyms and stories attached to him that the Bhutan landscape can be read as a map detailing his travels and activities. Serving as a marker of these activities are toponyms of the places that hold the stories, and beyond these stories, the history of the land. Thus, the oral hagiography, as narrated from place to place, manifests as a reading of the geo-ecological text of Bhutan. These stories contained in toponyms and cultural memory—in the form of oral accounts or written hagiographies—are referred to as *namthar* (རྣམ་ཐར་). Reading, listening, or even remembering these sublime stories engages and enables Bhutanese to sense and feel the text of the earth. Semantically, *namthar* is the telling of the most sublime story (Zangpo 2002). Embedded in the tones and textures of the landscape are important cultural signs and symbols that act as bearers of values. These are used to draw an understanding of our innate emplacement—physical as well as mental—with a thin line between material and mind, and place and space.

In a Story a Place is Born

As the previous section substantiated, in Bhutan, the enmeshment of narrative knowledge and the geophysical landscape is condensed in toponyms that embody the past, present, and future. Set in stories, toponyms are not just points of geographical reference but conduits of clime-making. As storied toponyms are passed down from one generation to the next, they become intangible heritage about lively tangible worlds. They enact and envision community by immortalizing social mores, religious virtues, and political wisdom in the land. The two ethnographic narratives that follow illustrate this further.

Haa, in western Bhutan, is a small town located between two mountain ranges. It is known as a *Baeyul*, which, I am told, invokes a hidden land, concealed due to its sacredness and sanctity by renowned Buddhist masters skilled in the ritual practice of hiding spiritual treasures. The afternoon sun is directly overhead, the sky is clear, as is the river that flows down the valley. I am talking to Ap Phub Tshering. Now in his eighties, he has lived his whole life next to the Lhakhang Karpo ("White Monastery").

"Haa is an unusual name," I say in the hope of prompting an explanation.

"It is an exclamation that became the name of this place," Ap Phub replies.

"Agay [an affectionate honorific for an elderly man], what do you mean that?" I ask.

He explains:

> It is said that Lhakhang Karpo is one of the 108 temples constructed by the thirty-third Tibetan King Songsten Gampo across the Himalayas in a single day. It is believed that the king did it to subdue an ogress residing in the Himalayan region. It is said that the king released two birds—one white and one black—toward the south of Tibet. He prophesied that wherever the birds landed, a temple would be built. This is where the white bird settled [pointing to the monastery on whose stairs we are sitting] and this temple is therefore known as the White Temple. People from across the Haa Valley gathered here to help with the construction. As they were working, they witnessed a group of people come out of the mountain to help to complete the task in a day. The hat on the statue and the pinnacle of the temple were offerings from the local deity Ap Chundu, who manifested as an old trader. Witnessing it in awe, people called the place "Hay Lung"—a valley where people miraculously appear out of nowhere for temple construction. The pronunciation later became Haa.

The storied toponym of Haa indicates how nonhuman entities, events, and places interact and shape one another.

From the Haa Valley, we climb up to the Bhutan highlands. "It must have been a tiring journey," Lopen Dorji, the vice-principal of the monastery, remarks. "But now that you are here, this place will make you forget your pain." It has taken me two days to walk from Gunitsawa, where the vehicular road ends, to the small herding village of Soe. Later, we meet Aum Tshering, the district administrator, who was born and raised in Soe.

"Would you mind welcoming me to Soe with a story about this place?" I ask. She replies:

> People often come to know about Soe because of the mountain Jomolhari. So, let me tell you a story about Jangothang, which is at the base of the mountain. Jangothang (བྱང་ཐོག་ཐང་) literally means a flat area with ruins of stone walls. You should go there tomorrow.

We immediately agree, and Aum Tshering continues with her story:

> Mounted on a huge boulder, you will see a ruin of a *dzong* [fortress], which was once the palace of a local king. Even today, it stands tall against the backdrop of the mountain Jomolhari. Below the ruins is a flat land triangulated by the confluence of two rivers—one flowing from Jomolhari and the other from the mountain Tshering Gang. It is said that, surrounded by these very high mountains, the king was unhappy as his palace hardly saw the morning sun during the winter. Therefore, he gathered his subjects to obliterate the surrounding mountains so the palace would receive the first rays of the sun each morning. Holding onto the Bhutanese dictum of "the king's order is like a waterfall that can never flow upward," his subjects started excavating the mighty

> mountains. One day, while scratching at the surface of one of the mountains, a woman sang, "Rather than attempting to obliterate the mighty mountain, it would be better to annihilate the mighty man." Following this advice, the king was killed and his subjects migrated out of Jangothang. Today, we believe that the king and the woman have been reborn as guardian spirits of the ruins.

Similar stories—all adapted to the local landscape—are told throughout Bhutan and in neighboring Arunachal Pradesh (see Gohain 2020; Chapter 10, this volume).

Aum Tshering concludes by saying: "I think this story teaches our age-old proverb: 'If you don't measure what you eat with a *drey*, every part of your body above the calves will become your stomach.'" (A *drey* is a traditional measuring cup that is used to weigh grain.) A symbolic wisdom about the evil of immoderate human desire stems from the storied toponym, as well as this adage. I understand the tale of the king's demise and subsequent rebirth as a guardian spirit as a warning of the dangers of greed, selfishness, and attachment.

These cases from Haa and Jangothang illustrate how stories make places, and how toponyms are carriers of historical events that subsequently turn in place-makers.

The Spiritual Instruction of the Landscape

In this section, we remain in Soe. As Ap Dawa Penjor pointed out earlier, a chain of mountains march to the Paro Valley. At Zombathang, as he explained, there is a uniquely shaped peak named Zangdog Pelri. Now, I am walking with Sonam Tenzin, who has been carrying out ethnographic research in the highlands for several months. As we reach Zombathang, Sonam stops, turns left off the trail, and points to a mountain. "Do you see that?" he says. The upper half of the mountain is cast in the morning mist. "The herders say that it is the representation of Zangdog Pelri."

As I stare at the majestic mountain, I see that it is indeed uniquely shaped, with dozens of equally shaped, spherical ridges that seem to be stacked in a perfect heap, reminiscent of *Rinchen Ph'ung-pa* (a Buddhist term for a heap of precious jewels). Sonam points out the copses of Himalayan larch at the base of the mountain and says, "The number of larch decreases, and the individual trees become weaker, higher up the mountain." As an environmental scientist, I think this is perfectly normal. In Bhutan, the treeline is about 3800 meters above sea level—far below the summit of Zangdog Pelri—so it is hardly surprising that the larch start to thin out at higher altitudes. However, Sonam offers a different explanation:

> The herders say that the larches at the base are people who have sinned gravely. Having committed these sins, they are reborn here and

> condemned to endless atonement by climbing the mountain. To atone for their sins completely, they have to move up to the ridge.

Few of the larch have advanced past the lower quarter of the mountain.

This penance of walking up Zangdog Pelri in tree form to mitigate the karma of past misdeeds reminds me of how Bhutanese engage in circumambulation. However, unlike humans, who are clearly on the move while circumambulating, the movement of trees is less obvious. Later, a herder tells me that the number of larch at the base has increased significantly over recent years, and that the trees are moving further up the mountainside each year. The local explanation for these changes is that more people are being reborn here, because more people are sinning during their human lifetimes. This seems an apt way of thinking about the human role in the Anthropocene, in which an unraveling earth is directly correlated with the sin of greed. It also echoes Wouters and Dema's account of the recent decline in auspicious *tshomen* sightings, which is explained in terms of younger generations' disinclination to cultivate reverential and caring relations with the physical earth (see Chapter 3, this volume).

My scientific study of climate change has taught me that the earth's higher temperatures allow trees to survive at higher altitudes. The herders of Soe have witnessed this shifting treeline at first hand, but their understanding of it goes deeper than scientific observation. To them, it is indicative of an accumulation of human sin, which is experienced and explained as the cause of climate change. While, as an environmental scientist, I would try to understand this in terms of numbers and models, the herders understand it qualitatively and relate it to their knowledge of Zangdog Pelri. This indicates how climate/clime change becomes incorporated, experienced, and explained in terms of pre-existing, emplaced knowledge. This is also demonstrated by Yangzom and Wouters (Chapter 10, this volume), who explain that herders are interpreting the unprecedented heat in the highlands as the realization of an age-old prophecy that foretells "immense heat from a combination of seven suns" that will lead to drought and conflict. The shifting treeline, as a representation of the accumulation of sin, similarly accounts for climate/clime change through embodied knowledge.

Conclusion

This chapter adopted a storying-climes approach to complement extant scientific understandings of Bhutan's landscape and climate with affective, lively, terrestrial, and multispecies accounts. Various ethnographic narratives reveal that, in Bhutan, the geophysical texture is enlivened by toponyms and stories that, in a sense, *perform* the worlds they narrate. These geomorphological textures become texts in a deep enmeshment between ecologies of biotic and abiotic entities, humans and their cultures, and deities in a long climatic, geological, evolutionary history. Toponyms, in particular, can be climed as a series of stories that variously affirm their role as clime-makers and as enabling an affective, relational, and multispecies reading of the Bhutanese landscape as a living geomorphological text.

Clime change is experienced in the threat to—and transformation and disappearance of—landscape components that are infused with spiritual, historical, and cultural meaning. Thus the melting of the glacial fortress of Ama Jomo, the flooding of the terrestrial deity Tsheringma's lake, the burning of a forest spirit's citadel, the shifting treeline on Zangdog Pelri, the drying of the *lu* (deities) of wet meadows, the changing color in Guru Rinpoche's copper abode. These are the ways in which clime change is experienced and explained in the Bhutan context. This makes clime change not only a climatological and ecological problem but also a cultural and religious encounter. It is in such facets that I identify the need to move beyond science-only explanations of climate science. We need more caring and capacious accounts of the ways in which multispecies communities in Bhutan, and across the Himalayas, experience, explain, and encounter anthropogenic transformations in earth systems and processes. The storying-climes approach employed in this chapter offers one such avenue to broaden our horizon of what clime/climate change is, and what it is about.

Note

1 The ethnographic dataset I use in this chapter was compiled during fieldwork conducted by a research collective that comprises Thinley Dema, Deki Yangzom, Jelle J.P. Wouters, and myself, aided by our research assistants: Sonam Tenzin, Pema Choden, and Kinley Wangmo. I express my gratitude to the entire team.

References

Allison, Elizabeth. 2019. "Deity Citadels: Sacred Sites of Bio-Cultural Resistance and Resilience in Bhutan." *Religions* 10(4): 268. https://doi.org/10.3390/rel10040268.

Banerjee, Milinda and Jelle J.P. Wouters. 2022. *Subaltern Studies 2.0: Being against the Capitalocene.* Chicago: Prickly Paradigm Press.

Basso, Keith H. 1996. *Wisdom Sits in Places: Landscape and Language among the Western Apache.* Albuquerque: University of New Mexico Press.

Chao, Sophie and Dion Enari. 2021. "Decolonising Climate Change: A Call for Beyond-Human Imaginaries and Knowledge Generation." *eTropic* 20(2): 32–54. https://doi.org/10.25120/etropic.20.2.2021.3796.

D'Avignon, Robyn. 2022. *A Ritual Geology: Gold and Subterranean Knowledge in Savanna West Africa.* Durham, NC: Duke University Press.

Gohain, Swargajyoti. 2020. *Imagined Geographies in the Indo-Tibetan Borderlands: Culture, Politics, Place.* Amsterdam: Amsterdam University Press.

Kimmerer, Robin Wall. 2013. *Braiding Sweetgrass: Indigenous Wisdom, Scientific Knowledge, and the Teachings of Plants.* Minneapolis: Milkweed Editions.

Phuntsho, Karma. 2013. *The History of Bhutan.* New Delhi: Random House.

Shepherd, Theodor G. and Chi H. Truong. 2023. "Storylining Climes." In *Storying Multipolar Climes of the Himalaya, Andes and Arctic: Anthropogenic Climate and*

Shapeshifting Watery Worlds, edited by Dan Smyer Yü and Jelle J.P. Wouters, 157–183 Abingdon and New York: Routledge.

Singh, Vandana. (2023). "Not Just the Science: A Transdisciplinary Pedagogy for Cryospheric Climes." In *Storying Multipolar Climes of the Himalaya, Andes and Arctic: Anthropogenic Climate and Shapeshifting Watery Worlds*, edited by Dan Smyer Yü and Jelle J.P. Wouters , 184–200. Abingdon and New York: Routledge.

Smyer Yü, Dan. 2023. "Multipolar Clime Studies of the Anthropocenic Himalaya, Andes and Arctic: An Introduction." In *Storying Multipolar Climes of the Himalaya, Andes, and Arctic: Anthropogenic Climate and Shapeshifting Watery Worlds*, edited by Dan Smyer Yü and Jelle J.P. Wouters, 1–26. Abingdon and New York: Routledge.

Smyer Yü, Dan, and Jelle J.P. Wouters, eds. 2023. *Storying Multipolar Climes of the Himalaya, Andes, and Arctic: Anthropogenic Climate and Shapeshifting Watery Worlds*. Abingdon and New York: Routledge.

Tilley, Christopher and Kate Cameron-Daum. 2017. *Anthropology of Landscape: The Extraordinary in the Ordinary*. London: UCL Press.

Tshering, Chang T. 2023. *A Collective History of Bhutan*. Thimphu: Centre for Bhutan and GNH Studies. [Title translated from the original Dzongkha.]

Ura, Karma. 2001. "Deities and Environment: A Four-Part Series." *Kuensel: Bhutan's National Newspaper*, November 26.

Wouters, Jelle J.P. 2023a. "Multilateral Clime Studies." In *Storying Multipolar Climes of the Himalaya, Andes and Arctic: Anthropogenic Climate and Shapeshifting Watery Worlds*, edited by Dan Smyer Yü and Jelle J.P. Wouters, 153–272. Abingdon and New York: Routledge.

Wouters, Jelle J.P. 2023b. "Where is the 'Geo'-political? More-than-Human Politics, Polities, and Poetics in the Bhutan Highlands." In *Capital and Ecology Developmentalism, Subjectivity and the Alternative Life-Worlds*, edited by Rakhee Bhattacharya and G. Amarjit Sharma, 181–202. Abingdon: Routledge.

Zangpo, Ngawang. 2002. *Guru Rinpoche: His Life and Times*. Boulder: Shambhala.

5 Climing Everest through Cryo-Visuals

Jolynna Sinanan

Introduction

Nepal has experienced two centuries of becoming iconic in "Western" imaginaries as an exotic Other based on images of the mystic and romantic appeal of the Himalayas and for narratives of heroism and spiritual fulfillment. Since the nineteenth century, the Himalayas have been scientifically, imaginatively, and politically remade to conform to an imperial order, which has had lasting consequences for the generation of knowledge about them (Fleetwood 2022). More recently, the global mediatization of Mount Everest (Chomolungma/Sagarmatha) has played a key role in attracting visitors to the region (Mazzolini 2015; Mu and Nepal 2016; Ortner 1999). Arguably, Everest has always been mediatized in contemporary global imaginations—its appeal as an idea has existed in part through technologies of visual cultures. It is especially mediatized now; recently improved digital infrastructure in the northern Himalayas has coincided with an increase in the number of tourists arriving between 2016 and 2018 (Ministry of Culture, Tourism, and Civil Aviation 2017). In turn, contemporary mediatization has had the effect of globalizing Everest, causing it to be predominantly seen and imagined according to colonial and global narratives that are decontextualized from the histories and cultures of the Himalayas.

This chapter proposes a cryo-climing approach that examines images of ice as communicative objects. Beyond being simply representations, these cryo-visuals seek to identify the values, narratives, and meanings attached to ice. Cryo-climing draws on the concept of climes, coined by James Rodger Fleming (2010, 7, 8) and developed by Mark Carey and Philip Garone (Carey and Garone 2014, 284, 285), to connect the local and the global, weather and place, nature and culture, and science and lived experiences to ground the more abstract use of "climate change." Clime represents place-based embodiment and an agent of climate change (Smyer Yü 2023, 8). However, in *Storying Multipolar Climes of the Himalaya, Andes, and the Arctic*, Karsten Paerregaard (2023, 53) innovates clime as a verb—"climing"—and as such captures climate change as processes of climing. Meanwhile, other chapters in that volume elaborate on climing to examine the processes and agencies

DOI: 10.4324/9781003484394-5

associated with "the actions of the physical Earth, the environmental flows, and humans and non-humans" (Smyer Yü 2023, 12). In this chapter, cryo-climing builds on these innovations to capture the social, ecological, historical, spiritual, and ritual relationships between humans and ice. Through the examination of images of ice taken over a period of time and by different stakeholders, cryo-climing enables more nuanced identification of the movement of ice, the movement and transformation of ice in relation to and as an indicator of climate change, and the relationship between verticality, ice, and climate change.

The enduring themes of spirituality, solitude, and adventure contribute to the appeal of Everest, attracting ever more tourists to the region, which is already fraught with anthropogenic alterations in both environment and climate. However, these dominant narratives fall short of capturing the totality of the complex relationship between climate change, development, and human and nonhuman agency in the region. The production and circulation of images through digital technologies shape how tourists imagine and experience Everest in Nepal. Drawing on fieldwork conducted in the Solu-khumbu region with guides, porters, and tourists, I argue that recently implemented digital infrastructures that facilitate mobile broadband and Wi-Fi on mobile devices (mostly inexpensive smartphones) may also provide possibilities for constructing alternative narratives of populations who either inhabit or spend large amounts of time in the region and whose livelihoods depend on increasingly fragile environments.

The chapter first establishes cryo-climing as a useful approach by presenting a brief history of the visual narrative influences that continue to appear in the ways that Everest is photographed and circulated, particularly on digital and social-media platforms. The imaginary of Everest evidenced in images of mountain- and ice-scapes echoes visual motifs of mountains accessed for exploration, scientific knowledge, and leisure as seen in Romantic paintings of the Alps in the eighteenth century. These narratives continue to influence the ways in which Everest is valued, consumed, and circulated. Such images are intended for global consumption, and as such do not reflect the same cultural, aesthetic, and religious experiences of the Sherpa and other populations connected to the region. At the same time, such images affect regional perceptions to varying degrees (Ortner 1999; Sherpa 2014).

The chapter then presents ethnographic vignettes to illustrate how these themes play out in the tourist encounter. Between 2017 and 2019, I conducted three visits of three to four weeks each to Kathmandu and the tourist hub of Namche on the Everest Base Camp trek. My visits in 2017 and 2018 comprised groundwork for fieldwork conducted in 2019. Further fieldwork was scheduled for the April–May summit season during Nepal's "Visit Nepal Year of Tourism 2020," but this was canceled due to the global COVID-19 pandemic. This was the second major disruption to Nepal and the Everest tourist industry in the space of just five years, after the earthquake and subsequent avalanches of 2015. Despite the high visibility of Khumbu Sherpa, who have been traditionally associated with Everest mountaineering, significant numbers of guides and porters are members of the Tamang and Rai ethnic groups, native to other parts of Nepal, who have historically been at the economic margins of Nepali society (Nepal 2005). Most of the workers who participated in this study were members of these groups who worked seasonally in the Khumbu region.

With Nepal's peak tourism season canceled and workers remaining in Kathmandu or in villages in the surrounding regions, the future of the tourism workforce in subsequent months remained uncertain. The chapter concludes by speculating on digital practices' potential to disrupt the aforementioned dominant narratives, and to provide alternative means by which other ways of knowing Everest ice may become valorized.

The tourist encounter reveals cryo-climing according to different stakeholders with unequal social relations of precarious labor where digital practices promote the commodification of fragile alpine ecosystems. On the other hand, they also reveal alternative narratives, values, and meanings based on the relationships of those whose livelihoods depend on mountainous environments. The chapter contributes to climing by examining the relationship between technologies and the human affective consciousness of water by considering the values of Everest's ice-scapes within local bounds and across global networks.

Cryo-climing: Knowing Everest Ice through Cryo-Visuals

Mount Everest, known as Chomolungma to Sherpa and Tibetans for centuries, has also been given the name Sagarmatha by the Nepali government, which founded the Sagarmatha National Park in 1976. Indeed, the exercise of measuring and "naming" Everest is one of subjecting the mountain to the imperial and geopolitical order of colonial control. In 1847, the Surveyor General of India, Colonel Andrew Waugh, reported to the Royal Geographical Society in London that his team had "discovered" the highest mountain in the world. Neglecting to mention any local name, he proposed calling it "Everest" after his predecessor, Colonel George Everest. A political officer in Darjeeling, B.H. Hodgson, pointed out that the locals already had a name for the mountain, but he was overruled so the "discovery" could be celebrated as a great achievement for British imperial science and knowledge (see Slemon 1998). This process of "naming" Everest reduced an affective, spiritual, material body to a topographical description in geographical and scientific terms. Consequently, it also reduced the meaning of the mountain within the context of the Himalayas, where it was understood to be a sentient being as well as home to an embodied and emplaced deity: the Buddhist goddess Miyolangsangma, who lives on the summit (Bernbaum 2022).

Today, Everest predominantly exists in global imaginations through technologies of visual cultures. Mass-produced and mass-circulated (through social-media platforms, for example) images reveal narratives that position humans as desiring, conquering, and colonizing Everest. Examining these images and the historicities they carry reveals the ways in which societies create and narrate ice as embedded in minds and identities, beyond its physicality (Sörlin and Dodds 2022). Focusing on the far north, Sörlin (2015) presents a brief history of the knowledge of Arctic ice. His "cryo-history" emphasizes the significance of communities and narratives: how ice is described, from which vantage point, and with what purpose (Sörlin 2015, 334). This chapter, in turn, focuses on Himalayan ice—an area that has received significantly less attention than Arctic ice—to develop a "cryo-climing" approach: through ethnographic inquiry coupled with historicizing ice in visual cultures, such an approach offers a broader historical reading of what can be thought of as "cryo-knowledge" that captures an ethos of better understanding of environmental change within global social processes for future orientation (van Dooren and Bird Rose 2016) or "cryo-narratives" that capture contemporary stories about ice (Leane 2017). The recent emergence of hydrohumanities and blue humanities, which consider discourses of water and power, and oceans, power, and history, aims to destabilize the colonial knowledge about human use and "ownership" of waterscapes (de Wolff and Faletti 2022). These turns in scholarship focus on water primarily in its liquid form. A cryo-climing approach offers an explicit focus on water's solid phase in the form of Himalayan ice. The complementary discussions between hydrohumanities, blue humanities and ice humanities thus retains a commitment to interrogate how scientific knowledge is generated and valued as well as whom it serves and excludes, and seeks to find and valorize alternative forms of knowledge that have been silenced, marginalized, or rendered invisible through colonial and imperial historical processes.

Sörlin and Dodds (2022, 8) argue that there are few studies on the visual representations of ice, and that ice in visual cultures captures only the surface value of ice, negating its qualities as an archive of the earth's dynamics. I propose a cryo-climing framework, exemplified through Everest in the multiplicity of visual cultures. Cryo-climing varies between polar and alpine regions and invites further, comparative discussion to avoid becoming a homogeneous approach and rather contribute to the rich forms of cryo-analysis where cryo-climing considers the connections between communities and places of ice (Cruikshank 2005; Leane 2017).

Cryo-climing considers the agency of ice and how it appears in visual cultures. Reflecting multipolar clime studies, it aims to unpack the multiple forms of knowledge that exist in addition to scientific models (Smyer Yü and Wouters 2023). Images are communicative objects that conceptually reflect cryo-climing: what kinds of knowledge and values do images of ice reveal? Elena Glasberg (2012) takes a similar approach in her analysis of Antarctica in photography and literature to critique the narrative lineage that produced the ways in which the continent is represented. An approach that draws attention to the social constructions of Everest invites us to lament the dominance of the Everest imaginary that positions the region as conquerable, and instead construct an approach where Everest may be seen, known, and valued for its place in a future where regional environmental degradation is reduced. By examining visual narratives of mountains and ice, such an approach addresses Kuntala Lahiri-Dutt's (2014, 507) call for thinking through the historical production of water/lands and the social construction of nature—with water, in this chapter, in its solid form.

Dan Smyer Yü (2021, 5) argues that research in the Himalayas that focuses on climate change and the environment is predominantly undertaken from the perspective of the natural sciences; as such, human experiences and narratives are rarely represented. The Greater Himalayas contain the largest mass of ice outside the polar regions, they are the source of the ten largest rivers in Asia, and they are the main water source for 22 percent of the global population (Xu et al. 2009, 972). Along with the polar regions, the Himalayas are experiencing climate change at twice the global average, warming at 0.3 degrees Celsius each decade (Davis et al. 2021). While the effects of climate change have become dominant narratives from techno-scientific perspectives, Chakraborty and Sherpa (2021, 51) call into question the construction of such representations by arguing that narratives of risk based on environmental determinism, state development policies, and an emphasis on techno-managerial decision-making contribute to the obfuscation of the plurality of worldviews in the region.

To counter these hegemonic views, Sherpa (2015) emphasizes the need to include perspectives from all the region's ethnic groups and to consider historical and migratory contexts. In addition, she insists that incorporating the views of residents will contribute to a better understanding of the short- and long-term effects of climate change (Sherpa 2014). The Solukhumbu region has already experienced the consequences of increased human activity through tourism, as seen on the Everest Base Camp treks, most obviously through the visibility of infrastructure and commercial activity between on-route and off-route villages. Sherpa (2014) notes that every on-route household is involved in tourism-based commercial activities, with some complementing this income with subsistence farming. Off-route households are less reliant on commercial activities related to tourism, but several have family members who work on-route or have migrated out of the region—to Kathmandu or abroad. Sherpa in diaspora have been recognized for continuing to engage with Everest through various cultural, spiritual, and affective expressions (Sherpa 2019). My fieldwork since 2017 is consistent with Pasang Yangjee Sherpa's findings, which show that, intergenerationally, younger members of the Sherpa population (those in their twenties and thirties) have migrated for study or work, resulting in an increase of workers from other parts of Nepal undertaking farming and commercial tourist activities—for example, as guides, porters, cooks, and managers of accommodation lodges.

In Nepal, the majority of the research conducted with populations in the Himalayas has been with Sherpa—the original inhabitants of the Solukhumbu. The region's cosmological and cultural meaning-making of climate is relatively well documented. Allison (2015), for example, describes Sherpa beliefs of deities inhabiting mountains and the negotiations that result from the economic opportunities of commercial climbing tourism. Sherry Ortner's (1999) long-term ethnographic study of Sherpa captures the simultaneously mutually dependent and culturally conflicting relationship between mountaineers and Sherpa in the late twentieth century.

However, within the last decade, although Sherpa populations continue to inhabit the Solukhumbu as farmers, owners of guest lodges, mountaineering guides, porters and other occupations related to the tourist industry, a significant number of seasonal workers in the region are members of the Tamang and Rai ethnic groups who have migrated to the Solukhumbu from their lower-altitude regions in search of more lucrative livelihoods. As these migrant workers are not indigenous to the region, their experiences and subjectivities, including perspectives of what their work and environmentalism mean within an entire worldview and its social relations, have been under-investigated. Workers who negotiate livelihoods, cosmological concerns, and the place of the environment within their worldviews face a compounded challenge, resulting in a geographic double bind. Tourism contributes to a situation of "can't win" with the incompatibility of economic growth and environmental sustainability (Bateson 1972; Eriksen 2018, 40).

A cryo-climing approach proposes to unpack the multiplicities of perspectives and values to identify the opportunities, challenges, ambivalences, and contradictions of navigating work, future aspirations, and environmental concerns for those whose livelihoods depend on the region. The significance of investigating the ways that populations visualize and narrate ice is captured by Nüsser and Baghel's (2014, 138) use of "cryoscapes," which they propose as "a conceptual framework to analyze the emergence of Himalayan glaciers in the context of a dynamic, globally imagined mediascape." They examine the ways in which Himalayan glaciers are known and imagined, pushing the understanding of glaciers beyond their physical properties, and incorporate non-human agency to recognize how different actors live with and alongside them. More widely, cryo-climing reflects Carey et al.'s (2016, 772) arguments for a feminist glaciology, which interrogates "how knowledge about glaciers is produced, circulated, and gains credibility and authority across time and space." This approach foregrounds knowledge about glaciers that has been marginalized or considered peripheral to traditional glaciology and interrogates who produces knowledge about glaciers, how knowledge about glaciers becomes credible, the relationship between scientific knowledge, power, and colonialism, and the potential of alternative representations (Carey et. al. 2016, 772). The distinct realities and implications for those who, to use Karine Gagné's term, are active stakeholders in "caring for glaciers" contributes to decentering the natural sciences and masculinist, techno-determinist narratives that have been produced historically (Gagné 2018; Carey et al. 2016).

Visual Cultures of Mountains and Ice: From the Age of Exploration to the Heroic Age

Klaus Dodds (2018, 79) argues that "to trace the human encounter with ice is to enter a world filled with cultural, folkloric, mythical, scientific and spiritual qualities." Since the Enlightenment, the conquest of ice and cold has been a narrative dominated by the image of European and North American men overcoming physical and mental adversity to master cold places, with those who survive attaining fame and fortune as renowned explorers or mountaineers (Dodds 2018, 53). These foregrounding narratives obscure the potential for the appreciation of ice as a critical holder of "earth memory," for example where scientific inquiry, policy-making, and public interest conceive of ice in its surface value only, to the detriment of cultural and social relevance and the value of ice-scapes such as glaciers to sustain entire ecosystems (de la Cadena 2015).

The ways that Everest appears when mediated through technologies of visual cultures is highly influenced by how mountains and ice were depicted in European Romantic and American landscape traditions, when nature was constructed according to the sublime and on notions of grandness and an emphasis on inner experience of what falls outside of reason and conventional understanding (Schaumann 2020, 8). The paintings of Casper David Friedrich (1774–1840) are perhaps the best-known European Romantic depictions of mountains and ice. Gordon Pignato (2017, 40) suggests that Friedrich "possibly perfected the representation of the sublime, conveying an emotional response to mountains that closely approximates the emotions travellers experience traversing alpine passes." Friedrich's use of lighting and perspective, which placed the viewer in a high location overlooking a small valley and large mountains in *Wanderer above the Sea of Fog* (1818; Figure 5.1), has infused tropes of more recent tourist photography, so much so that Sean Smith (2018, 181–182) identifies it as a common motif on Instagram (Figure 5.2). Smith describes the promontory gaze as replicating the themes of longing for the sublime and spiritual platitudes of both the natural world and the world within that are seen in European Romantic landscapes.

Arctic ice has been the subject of scientific investigation for several years; however, Himalayan ice has become subject to attention only more recently (see Chapter 1, this volume). Miyase Christensen and Annika Nilsson argue that a combination of distilling the satellite image data and communication through media using reader-friendly graphics "makes it easier to 'know' Arctic sea ice for the broader public" (Christensen and Nilsson 2017, 261). They draw attention to the complexities of multiple forms of data collection, technologies, and analysis used for scientific study and the role of media and communications in knowledge production about climate change. By doing so, they emphasize that the ways of knowing and seeing ice are strongly influenced by a "social chain of knowledge production" (Christensen and Nilsson 2017, 262). More recently, more-than-sensory perspectives highlight the entwined meanings of weather, place, and consciousness (Smyer Yü 2023, 11). Astrid Oberborbeck Anderson (2023, 79)

Figure 5.1 Caspar David Friedrich, *Wanderer above the Sea of Fog*, 1818, oil on canvas, 98.4 × 74.8 cm, Kunsthalle Hamburg, Hamburg, Germany
Source: Reproduced from Artble: www.artble.com/artists/caspar_david_friedrich/paintings/wanderer_above_the_sea_of_fog

highlights Inuit understandings of *sila* (weather) as having agentive qualities that influence daily practices. Such approaches appreciate the scientific knowledge of ice, generated by expertise, analysis, and technology, but also the mediation of narratives of ice by the historical contextualization of environmental change and the social values that appropriate such data, including indigenous knowledge traditions.

If the ice-scapes of the Arctic and Antarctic have been imagined as inaccessible destinations that can be seen and known only through technologies of visual cultures, then Everest are their more accessible counterpart. Everest—and Nepal more generally—has burgeoned as a tourist destination. It has been widely visited through different imaginings of the nineteenth, twentieth, and twenty-first centuries, ranging from aspirations

Figure 5.2 Trekker overlooking valley on the Everest Base Camp trek
Source: Used with permission from research participant

for spiritual pilgrimage to adventure travel, and there are now distinct expectations of what ordinary people can and should encounter: namely, solitude, spirituality, and mountain exploration (Liechty 2017). The global imagination of Everest is inextricable from processes of globalization that began in the late twentieth century. The contemporary imaginary of Everest is shaped by constructions of the mountain as "the commercialization of risk," "selling adventure," and "made for Hollywood disaster" (Palmer 2002), exemplified by the popularity of accounts of the 1996 disaster in Jon

Krakauer's (1997) book *Into Thin Air* and the IMAX film *Everest* (1998). As the vignettes in the next section highlight, these narratives continue to attract tourists to the region and compel guides, in particular, to negotiate between conforming experiences to meet the tourists' expectations and infusing their own visions of what tourist experiences should entail.

Imaging Everest Ice: Between Commercial Tourism and Reflective Livelihoods

Alina was in her mid-twenties and was spending three weeks in Nepal during her annual holiday from working in a financial consultancy firm in Moscow. She grew up in Sochi, where she recalled weekends and holidays in nearby coastal towns, skiing and hiking in the Caucasus Mountains. After arriving in Kathmandu she bought a ticket to the town of Lukla, "the gateway to Everest." Flights to Lukla are typically in high demand as schedules can be unpredictable due to poor weather at the high-altitude airport. Priority is usually given to passengers affiliated with organized trekking groups, yet Alina was determined to remain as independent as possible, rather than join a large, commercial expedition. Once in Lukla, she met Taral, who had extensive experience as a porter and moderate experience as an assistant guide. Alina wanted to embark on the "three passes trek," which is partially based on the Everest Base Camp trek—up to the acclimatization point at Dingboche—before diverging to a much more remote and rugged trekking route in the Solukhumbu. She explained,

> I'm not that interested in the Everest Base Camp trek itself. I did a lot of trekking, skiing, and mountain climbing in Russia, I've been to Elbrus, both the more luxury way and the more difficult trekking way, so I'm not so interested in the tourist treks. I read that the three passes [trek] is more beautiful, you can go by yourself, and you don't need as much time.

Alina wanted to work with Taral due to his familiarity with the region. Although she was confident in her map-reading ability and did not expect him to carry her pack, she did not want to be entirely on her own. He would be able to support their navigation and could assist with other aspects of the trip, such as taking photos of her with her iPhone.

By the end of her trek, Alina had countless wide-angle shots of ice-covered mountain landscapes as well as close-ups of river rocks, flowers, and prayer flags. Many of these images could be described as typical tourist photography: details by which to remember the trip as well as picturesque and idealized scenery for posterity (Urry and Larsen 2011). However, her Instagram posts and stories displayed visual tropes that are more typical of those shared on the site in the hope of attracting likes and followers. For instance, one was a "footie," showing her sneakers as she sat cross-legged against a picturesque ice-capped mountain backdrop (Haynes 2016). Another showed her posing on a ledge with mountains in the background (loosely reminiscent of the positioning of the viewer in Friedrich's mountain paintings). Alina spoke about consciously posting images that she considered attractive and spectacular. Although she was fond of her detailed images of the trek, whether behind the scenes in a guest house or at lower altitudes beside the Dudh Koshi (Milk River), she felt they did not conform to the type of photo that is popular on Instagram (see Figure 5.3). As a result, her preference for independent, self-reliant wilderness tourism, which emphasizes challenging oneself while minimizing one's impact (Turner 2002), rarely translated to her online posts, which instead reflected spectacular or unique travel experiences as a form of cultural capital for middle-class lifestyle aspirations (Polson 2016).

The norms and expectations of digital and social-media practices are also negotiated by Nepalis working in the tourist industry. Dawa was an experienced mountain guide, first for a British trekking company, then with a colleague who started his own small business in Kathmandu. By 2019, at the age of 39, he was in the process of setting up his own company in the hope that this would enable him to cater to clients who shared his eco-tourism vision. He explained that he did not want to run yet another Everest Base Camp trekking outfit in what was an already saturated market. His favorite region was Manaslu, which he described as more remote, more challenging, with a more "traditional" way of life. He hoped that visits to this region would encourage tourists to think more deeply about trekking in Nepal:

Figure 5.3 A selection of Instagram images tagged with #everest

> I would really like people to see other places in Nepal that are not so popular—like my hometown, Jiri. Jiri is very important in the history of Everest mountaineering. It used to be the start[ing] point on the trek before the Lukla airport was built; many guides and porters come from there, and hospitals and schools were built around the time that the British and Europeans were trying to reach the Everest summit for the first time. Today, many families still work in agriculture, the countryside is still very beautiful, and tourists could learn a lot from spending some time there. I would like my treks to include some home-stays—or, if tourists want more privacy, some accommodation in the countryside—[so they can] learn about how we live in Nepal.

Dawa accumulated the photos that illustrate his new company's website over many years of trekking with organized groups, and many of them could be termed typical tourist photography: ice-capped mountains and clients posing together in front of scenic landmarks. By contrast, the images on his Facebook page illustrate what goes on behind the scenes of experiences that are "staged" for tourists (Edensor 2001). For instance, some of the photos on his timeline focus on team workshops, such as first-aid and mountaineer training, safety demonstrations, and strategies for reducing environmental impact, while others show what is involved in setting up camp or clearing rubbish prior to leaving a site. These images may not be valued for their aesthetic qualities, but they provide an insight into the practices and concerns of those who work in the tourist industry and an alternative narrative to those that compel tourists to visit the region.

Dawa and many of his fellow guides are acutely aware of the environmental changes that have occurred over the last two decades. The path to the village of Gorek Shep provides a spectacular view of the Khumbu Glacier and icefall beyond Everest Base Camp. Dawa has several photos of clients posing in front of the glacier that reveal the extent to which it has receded over the years. His personal collection of images, which dates back to 2005, has grown considerably over the last decade due to the ubiquity of smartphone photography. In addition, the increasing availability of technologies such as GoPro and Wi-Fi mean that clients can now post images to social media almost in real time.

One of Dawa's colleagues, Sanil, lived in Kathmandu with his wife and daughter. Although his work as a guide for an international trekking company required him to spend long periods away from home, he was determined to remain active and informed within his community. He explained that he had taken several short courses in various Buddhist teachings and the history of Buddhism in Nepal, and he hoped to incorporate the ethics and philosophies he had learned, especially in relation to environmental and conservation values, within his work. He had recently succeeded in reducing the hours he was working for the international company, as the destinations it covered were proving less popular with tourists than the Solukhumbu. This gave him the freedom to organize trips for individual clients or small groups that better reflected his personal vision of the future of tourism in the region. First, when his clients arrived in Kathmandu, he would introduce them to places that were not on the large trekking companies' itineraries, such as Buddhist temples and sacred sites with historical significance. Here, he would explain the significance of the aesthetics in relation to cosmological beliefs and values, including the role of mountains and water. Later, on the trek, he would contextualize the natural landscapes they encountered and their relationship to the local population with reference to changes he had witnessed over the years, including the impact of tourism on livelihoods and its consequences for the environment. Sanil explained,

> I join bigger groups with the main [company] but I take personal clients where I suggest things on their itinerary they might be interested in as well as the things they want to see. I can take the time to explain to them more of the history and the culture of where they visit. Many clients understand that they are in the mountains but people live there as well, and they are interested in what this means for them and for nature … Plus, I am interested in history, and many clients like to know more. They seem happy to hear more about things like Buddhism and the importance of the mountains to Buddhist people.

As this brief vignette illustrates, Nepal's guides are actively constructing narratives that integrate their reflections on tourism with their worldviews, concerns, aspirations, and, increasingly, the experiences and livelihoods of people who live and work in the region. Such portraits offer a glimpse into the potential of a cryo-climing approach. More systematic collection, analysis, and archiving, enabled by changing infrastructural conditions in the region, will facilitate greater documentation and narration of ice-scapes in the Solukhumbu, where tourism is expected to have an ever-increasing impact on the region's environment and inhabitants.

Digital Infrastructures and Futures of Knowing Everest Ice

Since the avalanches of 2014 and 2015, the government of Nepal has made a concerted effort to increase telephone connectivity in the region. At the same time, private telecommunications corporations have invested in under-resourced and under-serviced areas in the Himalayas by providing digital connectivity through Wi-Fi that is accessible on mobile phones. Two mobile providers (NCell and Everest Link) now serve a region where, until 2000, the only form of communication was letter-writing. Residents of the Khumbu region can recharge their phone service through pay-as-you-go cards, and tourists can purchase a SIM card that is valid for a month as soon as they land in Tribhuvan Airport in Kathmandu. While the lucrative tourist industry has accelerated the demand for telecommunications alongside earthquake recovery efforts, it has also ushered in a new phase of communicating the imaginary of the Everest region, predominantly through images facilitated by newly established mobile media infrastructures. The expanding digital infrastructure also facilitates possibilities for the digital practice of workers and inhabitants of the region.

Carey et al. (2016) present a case-study of citizen science for glacier knowledge production in the Peruvian Andes. They argue that the combination of mountaineering and glaciology has yielded valuable information about glacial change and glacier-related hazards, with data collected from the mountaineering community, including guides and porters, complementing glaciological research (Carey et al. 2016, 59). As such, citizens and communities observing, reading, interpreting, explaining, and circulating anthropogenic changes in ice and glaciers in their everyday affective attachments and relations with their situated geo-ecological climes is a form of cryo-climing.

Although regional populations may not be scientific or technological experts, their data and environmental observations augment scientific knowledge because they are produced by populations who are intimately familiar with the region and whose livelihoods largely depend on it. Furthermore, their intergenerational relationships with the regional climate and its changes are largely understood through religious and cosmological lifeworlds (Haberman 2021). The brief examples presented above illustrate that the collection of observations and images in the Solukhumbu is already occurring and is likely to increase due to the expansion of mobile-media infrastructures. However, the value of the circulation of such images and observations as alternative narratives to the established ways that Everest ice is seen and known is yet to be established.

It is commonly known that tourism in the Solukhumbu has serious environmental consequences, especially in the form of pollution, as evidenced by the emphasis on rubbish and an increasing recognition of the impact of human activity and the presence of micro-plastics in the region's water sources (Bishop and Naumann 1996; Talukdar et al. 2023). In addition, due to Everest's ecological significance and location, the retreat of its glaciers has become an important indicator of the rate of climate change—a visual theme that appears consistently in before-and-after photographs (Garrard and Carey 2017); see Figure 5.4). Through recently implemented digital infrastructure in the region, workers in the Solukhumbu are well positioned to document and circulate perceptions and experiences of environmental change and the impact of tourism, and to capture nonhuman agency in dynamic geophysical processes (such as precipitation, ablation, and avalanches). Potentially, they are already cryo-climing through discussions with one another about the changes they observe each year. Their collective, longitudinal insights and experiences provide essential, meaningful, and situated knowledge for climate scientists in the region. For example, guides with over two decades' experience are reliable witnesses and reporters of glacial change, including melting. Their desire for secure, sustainable livelihoods for themselves and their families is negotiated alongside their values and worldviews, which are based on a form of environmentalism that is often inextricably linked to their cosmological beliefs.

Figure 5.4 Section of the Khumbu Glacier seen from the trek to Gorek Shep
Source: The author

Conclusion: Digital Media and Citizen Science as Cryo-Climing in the Solukhumbu

The chapter has contributed to climing by focusing on Himalayan ice and has argued that a cryo-climing approach has the potential to reveal the agency of ice as well as how inhabitants and workers in the Solukhumbu live with and alongside cryo-scapes. A comparative cryo-climing approach thus opens new pathways for examining regionally driven political ontologies of persons through (visual) practices (de la Cadena and Blaser 2018). I have drawn together a brief visual history of ice and mountains and their connotations with longstanding themes of desolation and exploration within the European Romantic painting tradition. By doing so, I have traced how nature–human relations have emphasized scientific knowledge (for colonial conquest) and the commodification of mountain- and ice-scapes historically and presently continue to pervade the Everest imaginary. Such narratives are apparent in the hundreds of thousands of images posted on Instagram and tagged #everest or #everestbasecamp, indicating the extent to which mobile phones broadcasting experiences of treks or summit expeditions have become normalized within Everest tourism. In turn, these narratives play a key role in shaping the value and perception of Mount Everest in the contemporary global imagination.

The chapter has also emphasized that newly implemented digital infrastructures in the Solukhumbu region facilitate possibilities for generating narratives that counter the dominant ways in which Everest (ice) is seen and valued. The local population and those whose livelihoods depend on spending large amounts of time in the region are already intimately familiar with the changes, challenges, and discursive responses to climate change alongside their own negotiations and practices. Moreover, as we saw in the cases of Dawa and Sanil, they are already documenting these developments and portraying individual and collective forms of meaning-making. Meanwhile, as Alina demonstrated, although tourist photography tends to emphasize visual tropes that conform to the popularity of certain genres on social-media platforms, there is some potential for tourists to disrupt the dominant narratives. The roles of governments and other major stakeholders have been beyond the scope of the chapter, but the future of knowing Everest ice and moving toward a discussion of overcoming the tensions between aspirations for economic growth and environmental sustainability in the region depends on these fraught, unequal relationships that play complementary roles in cryo-climing.

References

Allison, Elizabeth. 2015. "The Spiritual Significance of Glaciers in an Age of Climate Change." *Wiley Interdisciplinary Reviews: Climate Change* 6(5): 493–508.

Bateson, Gregory. 1972. *Steps to an Ecology of Mind.* New York: Chandler.

Bernbaum, Edwin. 2022. *Sacred Mountains of the World.* Cambridge: Cambridge University Press.

Bishop, Brent and Chris Naumann. 1996. "Mount Everest: Reclamation of the World's Highest Junkyard." *Mountain Research and Development* 16(3): 323–327.

Carey, Mark and Philip Garone. 2014. "Forum Introduction." *Environmental History* 19(2): 281–293.

Carey, Mark, Rodney Garrard, Courtney Cecale, Wouter Buytaert, Christian Huggel, and Mathias Vuille. 2016. "Climbing for Science and Ice: From Hans Kinzl and Mountaineering-Glaciology to Citizen Science in the Cordillera Blanca." *Revista de Glaciares y Ecosistemas de Montaña* 1: 1–14.

Carey, Mark, M. Jackson, Alessandro Antonello, and Jaclyn Rushing. 2016. "Glaciers, Gender, and Science: A Feminist Glaciology Framework for Global Environmental Change Research." *Progress in Human Geography* 40(6): 770–793.

Chakraborty, Ritodhi and Pasang Yangjee Sherpa. 2021. "From Climate Adaptation to Climate Justice: Critical Reflections on the IPCC and Himalayan Climate Knowledges." *Climatic Change* 167(3): 1–14.

Christensen, Miyase and Annika E. Nilsson. 2017. "Arctic Sea Ice and the Communication of Climate Change." *Popular Communication* 15(4): 249–268.

Cruikshank, Julie. 2005. *Do Glaciers Listen? Local Knowledge, Colonial Encounters, and Social Imagination.* Vancouver: UBC Press.

Davis, Alexander, Ruth Gamble, Gerald Roche, and Lauren Gawne. 2021. "Thawing Tensions in the Himalaya." www.lowyinstitute.org/the-interpreter/thawing-tensions-himalaya.

de la Cadena, Marisol. 2015. *Earth Beings: Ecologies of Practice across Andean Worlds*. Durham, NC: Duke University Press.

de la Cadena, Marisol and Mario Blaser, eds. 2018. *A World of Many Worlds*. Durham, NC: Duke University Press.

de Wolff, Kim and Rina C. Faletti. 2022. "Introduction: Hydrohumanities." In *Hydrohumanities: Water Discourse and Environmental Futures*, edited by Kim de Wolff, Rina C. Faletti, and Ignacio López-Calvo, 1–15. Oakland: University of California Press.

Dodds, Klaus. 2018. *Ice: Nature and Culture*. London: Reaktion Books.

Dodds, Klaus and Sverker Sörlin, eds. 2022. *Ice Humanities: Living, Working and Thinking in a Melting World*. Manchester: University of Manchester Press.

Edensor, Tim. 2001. "Performing Tourism, Staging Tourism: (Re)producing Tourist Space and Practice." *Tourist Studies* 1(1): 59–81.

Eriksen, Thomas Hyland. 2018. *Boomtown: Runaway Globalisation on the Queensland Coast*. London: Pluto Press.

Fleetwood, Lachlan. 2022. *Science on the Roof of the World: Empire and the Remaking of the Himalaya*. Cambridge: Cambridge University Press.

Fleming, James Rodger. 2010. *Fixing the Sky: The Checkered History of Weather and Climate Control*. New York: Columbia University Press.

Gagné, Karine. 2018. *Caring for Glaciers: Land, Animals, and Humanity in the Himalayas*. Seattle: University of Washington Press.

Garrard, Rodney and Mark Carey. 2017. "Beyond Images of Meting Ice: Hidden Histories of People, Place, and Time in Repeat Photography of Glaciers." In *Before-and-After Photography Histories and Contexts*, edited by Jordan Bear and Kate Palmer Albers, 101–122. Abingdon: Routledge.

Glasberg, Elena. 2012. *Antarctica as Cultural Critique: The Gendered Politics of Scientific Exploration and Climate Change*. New York: Palgrave Macmillan.

Haberman, David L. 2021. *Understanding Climate Change through Religious Lifeworlds*. Bloomington: Indiana University Press.

Haynes, Nell. 2016. *Social Media in Chile*. London: UCL Press.

Krakauer, Jon. 1997. *Into Thin Air*. New York: Villard.

Lahiri-Dutt, Kuntala. 2014. "Beyond the Water-Land Binary in Geography: Water/Lands of Bengal Re-visioning Hybridity." *ACME: An International Journal for Critical Geographies* 13(3): 505–529.

Leane, Elizabeth. 2017. "Fictionalizing Antarctica." In *The Handbook on the Politics of Antarctica*, edited by Klaus Dodds, Alan Hemmings, and Peder Roberts, 21–36. Cheltenham: Edward Elgar.

Liechty, Mark. 2017. *Far Out: Countercultural Seekers and the Tourist Encounter in Nepal*. Chicago: University of Chicago Press.

Mazzolini, Elizabeth. 2015. *The Everest Effect: Nature, Culture, Ideology*. Tuscaloosa: University of Alabama Press.

Ministry of Culture, Tourism, and Civil Aviation. 2017. *Nepal Tourism Statistics 2016*. Kathmandu: MOTCA.

Mu, Yang and Sanjay Nepal. 2016. "High Mountain Adventure Tourism: Trekkers' Perceptions of Risk and Death in Mt. Everest Region, Nepal." *Asia Pacific Journal of Tourism Research* 21(5): 500–511.

Nepal, Sanjay. 2005. "Tourism and Remote Mountain Settlements: Spatial and Temporal Development of Tourist Infrastructure in the Mt. Everest Region, Nepal." *Tourism Geographies* 7(2): 205–227.

Nüsser, Marcus and Ravi Baghel. 2014. "The Emergence of the Cryoscape: Contested Narratives of Himalayan Glacier Dynamics and Climate Change." In *Environmental and Climate Change in South and Southeast Asia*, edited by Barbara Schuler, 138–157. Leiden: Brill.

Oberborbeck Anderson, Astrid. 2023. "Climing the Andes: Vertical Complementarity, Transhuman Reciprocity, and Climate Change in the Peruvian Highland." In *Multipolar Climes of the Himalaya, Andes and Arctic: Anthroposcenic Climate and Shapeshifting Watery Lifeworlds*, edited by Dan Smyer Yü and Jelle J.P. Wouters, 52–68. Abingdon and New York: Routledge.

Ortner, Sherry B. 1999. *Life and Death on Mt. Everest: Sherpas and Himalayan Mountaineering*. Princeton: Princeton University Press.

Paerregaard, Karsten. 2023. "Pluriversal Tundra: Storying More-than-Human Ecologies across Deep, Accelerated, and Troubled Times." In *Multipolar Climes of the Himalaya, Andes and Arctic: Anthroposcenic Climate and Shapeshifting Watery Lifeworlds*, edited by Dan Smyer Yü and Jelle J.P. Wouters, 69–88. Abingdon and New York: Routledge.

Palmer, Catherine. 2002. "'Shit Happens': The Selling of Risk in Extreme Sport." *Australian Journal of Anthropology* 13(3): 323–336.

Pignato, Gordon. 2017. "The Intersection of Alpine Passes and Landscape Painting." In *The Mountains in Art History*, edited by Peter Mark, Peter Helmen, and Penny Snyder, n.p. Middletown: Wesleyan University Press.

Polson, Erica. 2016. *Privileged Mobilities: Professional Migration, Geo-social Media, and a New Global Middle Class*. New York: Peter Lang.

Schaumann, Caroline. 2020. *Peak Pursuits: The Emergence of Mountaineering in the Nineteenth Century*. New Haven: Yale University Press.

Sherpa, Pasang Yangjee. 2014. "Climate Change, Perceptions, and Social Heterogeneity in Pharak, Mount Everest Region of Nepal." *Human Organization* 73(2): 153–161.

Sherpa, Pasang Yangjee. 2015. "Institutional Climate Change Adaptation Efforts among the Sherpas of the Mount Everest Region, Nepal." In *Climate Change, Culture, and Economics: Anthropological Investigations*, edited by Donald C. Wood, 1–23. Bingley: Emerald Group Publishing Limited.

Sherpa, Pasang Yangjee. 2019. "Sustaining Sherpa Language and Culture in New York." *Book 2.0* 9(1–2): 19–29.

Slemon, Stephen. 1998. "Climbing Mount Everest: Postcolonialism in Culture of Ascent." *Canadian Literature* 158: 15–35.

Smith, Sean P. 2018. "Instagram Abroad: Performance, Consumption and Colonial Narrative in Tourism." *Postcolonial Studies* 21(2): 172–191.

Smyer Yü, Dan. 2021. "Situating Environmental Humanities in the New Himalayas: An Introduction." In *Environmental Humanities in the New Himalayas: Symbiotic Indigeneity, Commoning, Sustainability*, edited by Dan Smyer Yü and Erik de Maaker, 1–23. Abingdon and New York: Routledge.

Smyer Yü, Dan. 2023. "Multipolar Clime Studies of the Anthroposcenic Himalaya, Andes and Arctic: An Introduction." In *Multipolar Climes of the Himalaya, Andes and Arctic: Anthroposcenic Climate and Shapeshifting Watery Lifeworlds*, edited by Dan Smyer Yü and Jelle J.P. Wouters, 1–26. Abingdon and New York: Routledge.

Smyer Yü, Dan and Jelle J.P. Wouters, eds. 2023. *Multipolar Climes of the Himalaya, Andes and Arctic: Anthroposcenic Climate and Shapeshifting Watery Lifeworlds.* Abingdon and New York: Routledge.

Sörlin, Sverker. 2015. "Cryo-History: Narratives of Ice and the Emerging Arctic Humanities." In *The New Arctic*, edited by Birgitta Evengård, Joan Nymand Larsen, and Øyvind Paasche, 327–339. New York: Springer.

Sörlin, Sverker and Klaus Dodds. 2022. "Introduction." In *Ice Humanities: Living, Working and Thinking in a Melting World*, edited by Klaus Dodds and Sverker Sörlin, 1–34. Manchester: University of Manchester Press.

Talukdar, Avishek, Sayan Bhattacharya, Ajeya Bandyopadhyay, and Abhijit Dey. 2023. "Microplastic Pollution in the Himalayas: Occurrence, Distribution, Accumulation and Environmental Impacts." *Science of the Total Environment* 874: 162495.

Turner, James Morton. 2002. "From Woodcraft to 'Leave No Trace': Wilderness, Consumerism, and Environmentalism in Twentieth-Century America." *Environmental History* 7(3): 462–484.

Urry, John and Jonas Larsen. 2011. *The Tourist Gaze 3.0*. London: Sage.

van Dooren, Thom and Deborah Bird Rose. 2016. "Lively Ethnography Storying Animist Worlds." *Environmental Humanities* 8(1): 77–94.

Xu, Jianchu, R. Edward Grumbine, Arun Shrestha, Mats Eriksson, Xuefei Yang, Y.U. N. Wang, and Andreas Wilkes. 2009. "The Melting Himalayas: Cascading Effects of Climate Change on Water, Biodiversity, and Livelihoods." *Conservation Biology* 23(3): 520–530.

6 Dancing in the Rain

Climing the Monsoon in Pre-Modern Assam

Rima Kalita

Introduction

In June 2022, a government official out for a midday tea break grumbled about the scorching heat and longed for rain. Taking a sip of steaming milky tea, his face displayed unease over the bright sun. The tea seller lamented about decreased customer flow as a result of the recent flood in Assam. It had hardly been a week into the state's recovery from the year's second flood after two weeks of ceaseless downpours. Assam had been submerged in flood waters for days. Yet, amid the blistering heat, a longing for a fresh spell of rain persisted, and perhaps it always will. Such monsoon talks in everyday conversations reveal that humans in Assam have not perceived rain as a disruptive rival, but have rather danced in its rhythm. Humans have coexisted with timely and untimely erratic waves of torrents and annual flooding that have become integral parts of everyday chatter turned to cultural memories. There are abundant instances of this from the past. A seventeenth-century account tells how humans adopted measures indicating a negotiating relationship between themselves and water. In this account, an Assamese king made offerings of goats and ducks to resist the relentless monsoon-induced flood (Bhuyan 1933, 26). Similarly, in the eighteenth century, French visitor Jean-Baptiste Chevalier, who experienced a sky full of clouds, violent thunderstorms, extreme wind, and heavy downpours in the month of December, perceived Assam's weather as "unusual and unpredictable," in contrast to that of neighboring Bengal (Chevalier 2020, 164). Nevertheless, pre-modern Assam's "moody monsoon" appeared in varied narratives stored in a rich written and oral tradition.

This chapter expands the idea of clime-thinking by enlivening pre-modern literary sources on weather events and climate patterns in pre-modern Assam. It understands these more-than-monsoon "dramas" as a form of affective "climing." Inspired by Paerregaard (2023) and Brandshaug (2023), the focus is on climing water, and especially monsoonal water. However, unlike Paerregaard's (2023) idea of humans' social climing of water, this chapter shows the socio-cultural ways in which monsoon itself can be imagined and perceived, seen as to be climed from the perspective of a pre-modern world (in the eastern

DOI: 10.4324/9781003484394-6

Himalayan flatlands). Simultaneously, climing here refers to a set of more-than-human practices that relate and affirm the monsoon as a lively agent of the weather system. Naming this "monsoon climing," the chapter explores the terrestrial understandings of the relationship between the monsoon and the physical environment that compels humans and other-than-humans to make adaptive responses in material, cultural, and spiritual terms. In Assam, monsoon climing entails being attuned to the dynamics and changes in the monsoon and related weather patterns, and also involves rituals and other forms of communication and commensality between the monsoon and the region's humans. This climing approach allows me to elucidate the various roles the monsoon plays as a warrior, a savior who looks after boats in the water, a lover, and sometimes a friend whose presence or absence alters one's mood.

The chapter also applies the idea of a spatio-temporal identity of a particular place or region corresponding to its distinct monsoon clime environs. It attempts to bring back the idea of weather as region—for instance, the "monsoon clime" of the easternmost Himalayan flatlands. Here, rain, as a sovereign over all, took charge for more than seven months in the pre-modern period (primarily April to October) (McCosh 1837, 71–72). This wet climatic zone across the eastern Himalayan foothills ruled over by a long monsoon is hence reassessed as a monsoon clime in this chapter. Thinking with clime opens up myriad meanings, ideas, and propositions. In particular, it offers a fresh avenue for cultural, relational, terrestrial, and multispecies understandings of climate. However, the ideas around climate can be diverse and historically shifting (Hulme 2017). The term "clime" denotes multilinear dialogues between environmental "seens" and "unseens" (for instance, between places and weather) (Carey and Garone 2014, 284). Attached to the fields of climate study, geography, and history, clime is an open-ended understanding that encapsulates a range of varying interlinkages, such as the entangled, animated, spirited relationships between humans and nonhumans, land and water, weather and lore, biophysical entities and art, and so on (see Chapter 1, this volume). Within the emerging field of clime studies, this chapter then leads readers to pre-modern Assam, a Himalayan flatland, to think afresh about the sensory and imaginative entanglements between monsoons and places, people, plants, and animals. The ingenious monsoonal drama's plot moves through the pre-monsoon to post-monsoon variables that are never constant. Attuned to that, this monsoon clime receives torrential showers in three phases—pre-monsoon, monsoon, and post-monsoon (Pradhan, Singh, and Singh 2019). Consequently, the long monsoonal showers and the rainwater discharge braid together to create and sustain new lives as well as eco-social and cultural beliefs.

Assam's historical monsoon has increasingly received scholarly attention that has brought the water to life as a protagonist in the making of the region (Saikia 2019). Adopting a *longue durée* approach, Saikia discusses how the Brahmaputra River shaped Assam both terrestrially and spatially, from sediments and alluvial deposits to the expansion of agriculture, water engineering, fish transmigration, and varied cultures. In his work, the Brahmaputra, as an antecedent of all, navigated the history of a major river valley in the eastern Himalayan flatlands. Complementing this river approach with an intimately linked monsoon approach, this chapter discusses how the monsoon plays the role of protagonist in this clime's pre-modern history (1500 to 1800 CE); how monsoon was perceived; and what it signified to the humans of the region. It explores the clime concept primarily through pre-modern literary productions to understand how lives in the clime registered climate variables (the timely and untimely rainfall patterns) and coexisted in the given temporal frame.

Monsoon has been a force of the environmental shaping of this fluid, unruly clime. Often, the clime's humans' concomitant engagements with interpreting the functions of weather, rain, or clouds were shaped by rainfall patterns. Recurrent physiographical and geomorphologic transformations, predominantly caused by heavy rainfall for eight months and resulting annual inundation, the constantly changing nature of the river and its banks, were the major co-designers of the lives herein. Aided by heavy rainfall throughout the year, the Brahmaputra and Barak river systems wander, shaping the clime's landscape into two major valleys. The terrestrial space of the Brahmaputra Valley, which spreads along the foothills of the eastern Himalayan mountain range, is a hub of human diversity in terms of social formations and cultural practices shaped primarily by the ecology.

Drawing on the rich clime archives that were byproducts of socio-cultural and intellectual activities in the Brahmaputra Valley, this chapter explains the importance of the study of a monsoon clime. Equal importance is assigned to climate archives to understand the meteorological, terrestrial, and environmental events of the monsoon. While reliable proxy data were mostly recorded only from the nineteenth century onwards, one cannot overlook the monsoon histories found in the pages of literature, rhetoric, and embedded histories, despite their thin presence. It is true that historians should be in constant dialogue with the present quantitative method of climate variability and meteorology to be able to understand the pre-modern climatic environs (Ladurie 1971). In another context, historian Geoffrey Parker (2013) highlights the importance of alternative methodologies for extrapolating future climatic predictions with quantitative data, and how one can "rewind the tape of history" with those nuanced observations. He advises studying the genesis, impact, and consequences of past catastrophes, using two distinct categories of proxy data: a "natural archive" and a "human archive." Both the first category—which comprises ice cores and glaciology, palynology (the study of pollen), dendrochronology, and observation of speleothems (cave structures), and the second—which comprises narratives, pictorial representations, archaeological shreds of evidence, and instrumental data, are essential tools to explore varied archives for the discourse of environmental humanities. Parker's work has been instrumental in helping me understand the nuanced monsoonal lives and their climatic variations and expressions. This chapter relies on his proposition to explore how understanding Assam's monsoon clime in its totality can greatly benefit from a critical dependence on pre-modern historical and literary productions. Hence, its clime understanding is drawn from the archives of various social groups, primarily the Ahoms, the Koches, and the Kacharies. The chapter engages with several literary genres, including various folk and oral traditions, Ahom royal chronicles (*buranji*s), published and unpublished Vaishnavite literature and hagiographies, and translated Mughal sources in both illustrated and non-illustrated forms. In addition, it draws a certain amount of historical understanding from later colonial publications.

Climate historians often rely on textual sources to understand periods of extreme weather, fluctuations in harvests, prices, and mortality, and so on (McNeill 2003); see, for instance, the fifteenth-century European witch trial reports (Carey and Garone 2014), the effects of the Little Ice Age (1550–1850 CE) that impacted the pre-modern mental world of Europe (Behringer 1999), or discussions on climate as a crucial component of cultural transmission (Bristow and Ford 2016). Answers to ambiguous questions on climate are examined by looking at micro-regional indigenous understandings of the seasons, the recurrence of patterns, and weather oscillations. Equally important is to understand spatio-temporality in relation to "deep climate" to uncover the evolutionary process of a clime. Clime not only bears the meaning of physiographical changes along with the temporal, but should also be considered as the "rhythm" between humans and more-than-humans, and between temporality and materiality (Irvine 2020).

Climing the Savior Monsoon

Monsoon is the product of the occasional heating and cooling of land and ocean. Terrestrially, it is a shaper of agentic entanglements between places that include humans and other-than-humans. But it has long been claimed that the monsoon's perpetual presence, which affects every part of the biome, is parallel to a drama, with multiple roles played by a single actor. Affecting half of the world's population, the Indian monsoon stages this monologue from Tibet to the southern part of the Indian Ocean, and from eastern Africa to Malaysia (O'Hare 1997, 218; Cook et al. 2010, 486). In these Asian–Indian monsoonal dramas (patterns), the eastern Himalayan foothills receive two waves of monsoon—in summer from the southwest and in winter from the northeast. The story of this monsoon monologue begins between these two waves.

Before the actual monsoonal drama sets in, the pulsating performance of the windy and stormy actors is crucial in the making of this clime; at the commencement of the monsoon, the pre-monsoon westerly wind changes the weather system. Sometimes, the clime experiences devastating effects during the months of March and April, as westerly winds accompanied by hailstorms damage crops, injure animals and birds, and lead to widespread destruction (Neog 1987, 192). These pre-monsoonal westerly winds sound the alarm for the arrival of the showers and prepare humans and nonhumans for monsoonal engagements.

A wide range of literary sources from the seventeenth century paint the monsoon as a savior on the battleground of the Ahom–Mughal wars, when the water transformed itself into a "knight in watery armor." Accounts from that era tell how the monsoon navigated the course of these wars in the seventeenth century.[1] Human warriors followed the "knight," frequently demonstrating their adept control over water. Mughal chroniclers applauded the Assamese for their naval expertise. The local king (of the Ahoms) strategically retreated to the uplands before the monsoon's arrival, but then resumed battle as soon as the heavy downpours began (Blochmann 1872, 79). Battling over and through the water turned it into a weapon itself, or indeed a savior (Nathan 1936, 611, 673). Well acquainted with the dynamics of the monsoon, the Assamese took advantage of the situation by disrupting their opponents' supplies and transportation (Gait 1906, 134–136). The peak of the rainy season was a suitable time to launch attacks, aided by the monsoon and inflated rivers. Using the monsoon as a front brigade, battle strategies were designed in the watery landscape of the Brahmaputra. The combination of might and monsoon supported the guerrilla warfare of the clime. Yet, over time, the waterscape shaped by long monsoonal cycles subverted the rhetoric of human might, as the monsoon clime shielded the Assamese from Mughal territorial ambition.

The monsoon savior also actively regulated the clime's accessibility. Its onset either permitted or prohibited movement for both humans and nonhumans. Sankardeva, a sixteenth-century Hindu saint, religious reformer, and literatus, poetically described how movement was restricted in the rainy season. He drew a dramatic analogy by comparing abandoned roads during the monsoon with a knowledgeable man abandoning reading (Dutta Baruah 1953, 66). Participating in the monsoonal drama, Assam's rivers played the catalyzing role as intermediary by holding and releasing rainwater during the monsoon. The Brahmaputra acted as the main supporting character in climing the monsoonal drama inherently entangled with polity, economy, and culture (Coleman 1969; Saikia 2019). Especially during the monsoon, the river in full spate transported fish and minerals (including alluvium) and, most importantly, discharged rainwater, giving impetus to the growth of rice culture in India's northeast. And the monsoon-fed river remained a savior, despite its phases of disruption. The river (and its environmental dramas) also poetically emerges as a mythical hero. An Assamese verse dating back to 1807 that describes the flooding caused by the Brahmaputra also tells how it devoured land belonging to a monastery as an act of its own divine cleansing.[2] The river sometimes aids the monsoonal dramas and memories with multiple meanings.

Monsoon stages its drama by taking the lead in the lives of both humans and nonhumans attuned with the rainfall and its distribution through rivers, rivulets, canals, and manmade reservoirs. Considering the presence of water and rain, human settlements were selectively situated in the fertile floodplains (*chapori*) along the riverbanks (Saikia 2019). But navigating lives in such a clime was not without its challenges. Exposure to the damp landmass and excessive rainfall inspired the design of the high-raised houses. Along with the human inhabitants, these houses also accommodated nonhumans (mostly domesticated pigs and goats) underneath the high platforms (Peal 1882, 53–55). Such monsoon-induced designs certainly reduced the distances between the worlds of humans and nonhumans and built monsoonal lives that were in sync with each other.

The damp, loamy soil of the extensive grasslands enabled large herbivores like elephants and rhinoceros to thrive, which in turn facilitated the coexistence of carnivorous species. The survival of the rhinoceros suggests no monsoon failure in the clime as they mostly feed on tall elephant grasses that can grow only in soft, loamy soils (Bose 2020), supported by the monsoon. However, anthropogenic changes of the Ganges Valley, chiefly through agriculture and hunting, caused the retreat of the rhinoceros from that landscape (Rookmaaker 1999; Bose 2020). The monsoonal clime in the Brahmaputra valley, on the other hand, actively hosted the verdant plant world that thrived through regular consumption of nutrients from the damp ground. For plants used in agriculture, cropping patterns that reveal the spectacular variety of rice grown were consistent, reflecting regular hydrological cycles. Typically, the whole of Southeast Asia and Assam, which are in close proximity to the eastern Himalayan ranges, the Tibetan Plateau, and most importantly the Bay of Bengal, relied on the southwest monsoon for agricultural activities. In addition, the southwest monsoon-induced wind circulation curated the seasonal cycles, staging a cool and dry winter and a warm and wet summer (Mehrotra et al. 2014, 42), which suit the cropping cycles. These two supporting actors—the Aeolian influence and alluvial sediments transported through wind and water flow facilitated by the Brahmaputra and its river system—constructed this monsoon clime's complex watery niches in support of the agricultural patterns. Favored by the long monsoon showers, rice had become the dominant staple crop in the Brahmaputra Valley by the sixteenth century (Ludden 2005), whereas millet fulfilled that function in upland India (Parker 2013, 98). Three principal rice varieties—*ahu* (dry variety), *bao*, and *sali* (long-maturing wet rice)—became supporting actors in transforming the region into a rice zone in the pre-modern era. These were grown solely in conjunction with monsoon patterns.

Monsoon and annual flooding intimately integrated cropping and the rainfall patterns. The humans of this clime attempted to coexist with the monsoon flood by sowing two different varieties of rice (*bao* and *ahu*) in the same paddy field (Saikia 2019, 252). Equally, there were attempts to build rudimentary dykes and embankments to control excessive floodwaters, though the people were more attuned to adaptive measures. Knowledge of how to control floodwater was evidently widespread. In the sixteenth century, a woman from a so-called "lower caste" took charge of managing the whims of excessive water flow by building a bund (Neog 1987, 184–185). Humans climed all the phases of the monsoonal drama in such a way that every actor in it intricately shaped their food habits to their social worlds.

Post-monsoon, in November–December, rice harvesting became a cause for social celebration climed through a harvest festival (*Maagh* bihu).[3] Many folk tales elaborately illustrate the monsoonal drama and its hydrological cycles in connection to these harvest—and feasting—festivals. In one such tale, a tiger and a crab celebrate the post-monsoon feasting festival (Bezbaroa 1988, 862–863). Aligning with the human imagination of animals creating farmlands, reverence for the harvest bears the symbolic meaning of nonhumans' equal engagement in the monsoonal drama. Climing the eco-cultural and religious customs together, humans of this monsoon clime celebrated a wide variety of festivals—*Roa-gara Puja, Lokhi Dak, Gosor Pana, Dhan-kaata*, and *Naya-khowa Puja*—to honor the paddy fields. Revolving around the cropping cycle, from sowing seeds to harvesting and finally consuming the rice, the *Roa-gara Puja* was conducted with particular enthusiasm in anticipation of sufficient rain for the summer crops (Sanyal 1965, 139−143).

Apart from staples, other seasonal crops, including pulses, fruits, herbs, and cash crops, were creatively cultivated and grown so that failure of one crop or grain never led to dire consequences. Mughal sources extensively documented this clime as a fertile and richly cultivated delta with remarkable vegetation, including natural orchards of tropical fruits (Vansittart 1785, 461–463; Blochman 1872). The regular monsoon cycles accommodated the farming of new, imported crops, including tobacco, which was introduced to Assam along with chillies and pineapples by the turn of the seventeenth century. The cultivation of poppies—another rain-fed cash crop—brought further economic effects from 1770 onwards (Blochmann 1872, 77, 91; Guha 1991, 85, 141). Additionally, fish, a crucial source of protein for humans, relied on the monsoon. The dietary requirements of the multispecies ecology of this clime entirely depended upon the long monsoonal dramas that intricately entangled the lives therein, with nuanced variations at times.

In addition to climing the role of the savior in the multispecies food ecology, the monsoon actively mediated lives between the uplands and the lowlands. A good deal of mercantile, political, and social activity depended upon the monsoonal flows (Ludden 2005). In the folk tale "The Kite's Daughter," a trader from a distant land meets the protagonist after taking refuge under a tree to escape the scorching heat (Bezbaroa 1988, 877). Merchants in such folk tales are often portrayed as fathers or husbands who spend most of their time away from home owing to the long monsoon. Similarly, in pre-modern hagiographic literary records, traveling to distant places is conventionally associated with the monsoon-fed rivers (Neog 1987; Neog 1950; Nath 1998). Rivers (including the Brahmaputra) were allegedly "bottomless" in the month of December (Blochmann 1872, 65). To bypass arduous land routes, humans made full use of convenient waterways, including rivulets (*nalah*), which were swelled by monsoon water at certain times of the year. Such narratives indicate the existence of active and dynamic communication with distant places, aided by the monsoon.

During the monsoon, the Brahmaputra, which descends into Assam and Bengal after meandering through Tibet, served as an important highway to large South East Asian markets. In other words, the rainwater that fed the river endowed the clime's economy with a potent trade network that extended from the eastern part of the monsoon clime to upper Burma and on to southwest China. This network was maintained across the hills long before the pre-modern era. Later, from the sixteenth century to the second half of the eighteenth century (1770), in the wake of political and economic expansion, salt and saltpeter were increasingly imported from Mughal India in return for forest goods, mustard seeds, and inferior-quality gold (Guha 1991, 1, 7, 85). Other products, such as *muga* (*Antheraea assamensis*) silk, found markets on the Coromandel and Malabar coasts via Bengal (Guha 1982, 490–491). In this conjuncture, the humans of the monsoon clime set out in their boats across the Jaintia Hills to Sylhet, from the Garo Hills to Bhutan, and up to Bengal in the west (Hamilton 1940, 76; Neog 1987, 214, 217, 249; Guha 1982, 488). These connected water routes enabled the transformation of the monsoon clime into a commercial hub. By utilizing the local biological resources, humans took part in the emerging global markets that intensified from the sixteenth century onward. Geopolitically located as a mediating region between Southeast and Central Asia, the clime's resources were in high demand in markets outside the valley. Supported by their monsoon savior, both upland and lowland humans positioned themselves socially according to their economic growth.

Monsoon, Rain, and the Lovers

Whether unreciprocated or reciprocated, water, the monsoon, and rainfall have often been climed as playful, romantic, and spiritual ideals in the literary and cultural imaginations of this clime. For instance, in popular folk narratives, they appear as lively invigorators of the mundane world, such as *Dighal Thengiya*, a personification of rain with unusually elongated legs (Bezbaroa 1988, 870–871). As the commencement of rain heralds the onset of three distinct seasonal transitions−spring, summer, and winter—spring festivals like *Bohag* bihu, *Bwisagu*, and others commemorate love, energy, and the well-being of the ecosystem. The drama unfolds with the arrival of the monsoon and ends with its departure. The monsoon's amorous play has been an important motif of union and separation in love. Sankardeva, who initiated the pre-modern genre of *bhakti* literature, described the exquisiteness of the monsoon as an aesthetic occasion—a dramatic excuse for connecting water bodies with human and aquatic lives (Dutta Baruah 1953, 65–68). In a long prose piece on the subject of rain, *Varsha Varnan* (*Description of Rain*), he romanticizes the monsoon. In Sankardeva's imagination, the monsoon bug *indragopa* (*Trombidium grandissimum*, or red velvet mite) transforms into a caterpillar (*guwali polu*) that is local to this clime. The anonymous artist of this illustrated manuscript filled the landscape of Vrindavana (a city in northern India where the Hindu god Krishna was believed to have spent his childhood) with the dramatic visual imagery of birds and cattle expressed in regional idioms.[4] The monsoon and narratives of love became the ingredients for imagery that could express many moods. Since there are no facial expressions to evoke emotions in the paintings, the presence of trans-representational nonhumans is crucial to understanding the moods of particular seasons (Figure 6.1). These nonhumans are depicted in carefree poses, frolicking to music played by Krishna, with their more-than-monsoon engagements becoming part of the entire monsoon drama. Other representations of the monsoon feature neither humans nor animals. For instance, in his most celebrated work, *Kirtan*, Sankardeva describes a luxuriant monsoon landscape that is full of love. There, the poet sets an other-worldly scene, portraying pristine nature with 44 plant species (Dutta Baruah 1953, 47–48).[5] He and later painters lyrically represented how they personally witnessed, and sensed, the monsoon. These visual representations indicate the enduring engagements of humans with constant monsoonal dramas played out in the sky. Although the painting tradition of this clime depicted classical narrations, the style equally embraced folk tales. The depiction of the monsoon's female lovers (local women) enthralled by monsoonal bliss blended with devotion captures the quintessential romantic mood (Figure 6.2).

Figure 6.1 Illustrated folio from *Chitra Bhagavata* depicting a monsoon-scape in Vrindavana with a number of nonhumans
Source: Dutta Baruah 1950

Figure 6.2 Illustrated folio from *Chitra Bhagavata* depicting a monsoon-scape in Vrindavana with the iconographic depiction of local women in the extreme left corner
Source: Dutta Baruah 1950

In terms of performative expression, a genre of seasonal songs—the *baramaahi* tradition (*barahmasa* or songs of twelve months)—depicts complex emotions that echo the different seasons (Gogoi 1991, 156). Considered as songs of separation, this genre of pre-modern lyrical poetry narrated the passing of the months and the moods of the seasons in terms of deep personal feelings (Pollock 2006, 144). Common symbols in pan-Indian *barahmasa* songs include: rain as a symbol of emotion; cloud cover as a symbol of the (male) beloved; and monsoon birds such as *chataks* (peacocks) as symbols of women transfixed by love (Vaudeville 1986). From this genre, the *Beula Baramahi Geet* illustrates pre-monsoonal hailstorms and heavy rain during the spring. It lyrically climes fresh rain showers' responsibility for spring blossoms and portrays a distressed heroine who is reminded of her long-lost love. Setting up the drama of love and pain (*viraha*), these songs sometimes express the reactivation of unrequited love during the monsoon. Conversely, the travelogues of occasional visitors climed the monsoonal drama as an eerily unhealthy unsolicited lover. For instance, Chevalier (2020; written 1755) accused the monsoon of causing high fever, loss of appetite, and eventually death. Such ailments were routinely attributed to the excessively cold water that descended from the mountains. Chevalier (2020, 134, 164) also bemoaned the monsoon's lashing rain, strong winds, and toxic air. Similarly, a Persian chronicler climed "this lover" as a mysteriously unhealthy someone/something, especially for visitors, documenting various complaints, such as nauseous diseases and unnamed epidemics (Blochmann 1872, 77, 88). The pre-modern monsoonal drama is found in popular narratives according to the imagination of the narrators. Whether as the synonym for a mysterious rainscape or the reason for blossoming love, monsoon as protagonist persisted as part of real and imagined perceptions.

Visualizing Deluge and Drought in a Monsoon Clime

The monsoonal drama reaches its climax in scenes of deluge and drought. This monsoon clime is evidently part of the broader Asian weather system. The Asian monsoon is particularly renowned for its complex spatio-temporal variability (see Cook et al. 2010). Evidence from earlier centuries confirms such variations. It is now well recognized that climatic change from tropical to subtropical monsoonal phases can be traced back to the end of the Pleistocene era (Hazarika 2017, 27). Recent speleothem records from Meghalaya reveal a major irregularity in monsoonal precipitation in the northeast region between the sixteenth and nineteenth centuries CE. Paleoclimatic records dating between 1710 and 1800 reveal a high frequency of strong and weak monsoonal shifts that signaled an untimely precipitation pattern (Gupta et al. 2019). Historical records do not indicate droughts, tornadoes, or cyclones, although this clime was a potential hotspot for all.

However, ceaseless rain for more than seven months each year and its pervasive presence from pre-monsoon to post-monsoon were part of everyday pre-modern life. In Sankardeva's *Chitra Bhagavata*, pre-modern artists (*khanikars*) documented the aesthetics of rain, personified gray clouds as demons, and painted raindrops as countless yellow dots (Figure 6.3). The outrageously dramatic depiction of demonic clouds reflects an understanding of the monsoon that is similar to that found in traditional one-act plays (*bhaona*).

Climing the drama of annual inundation was reflected visually in the sixteenth-century *bhakti* text *Anadi Patana*, a work devoted solely to environmental imagination (Figure 6.4). Illustrated in the eighteenth century, this cosmological text graphically displays how the monsoon and rain catalyzed a cosmic deluge

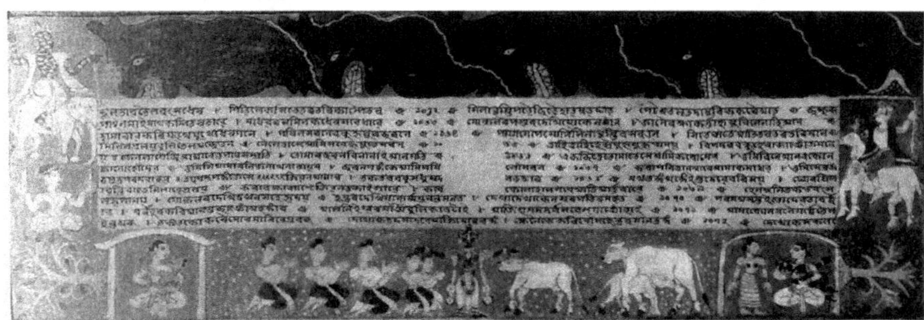

Figure 6.3 Illustrated folio from *Chitra Bhagavata* depicting humans and other-than-humans in a hailstorm in the Braj region
Source: Anonymous n.d.

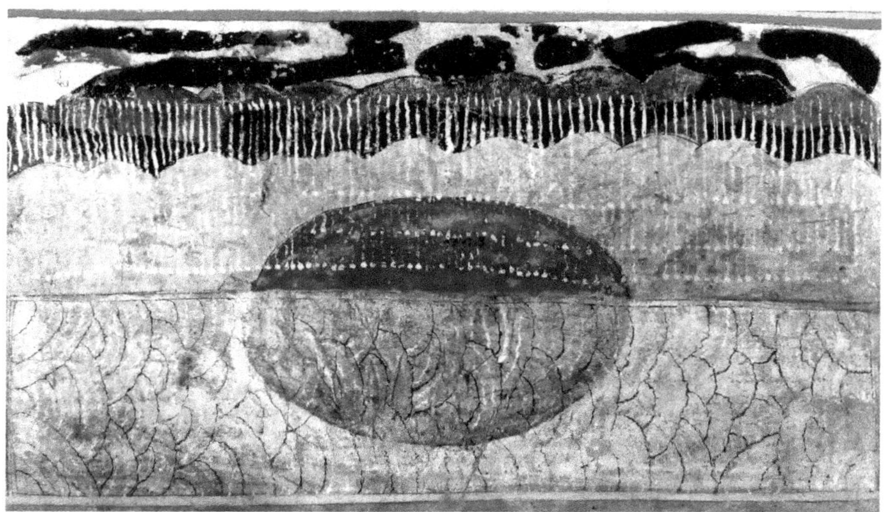

Figure 6.4 Eighteenth-century illustrated folio from *Anadi Patana* featuring a visual representation of "the Great Deluge"
Source: Kalita 2013

that plays an integral part in many creation myths. Here, the most remarkable performance of rain is reproduced visually in the form of numerous white lines descending from the black, tinged-with-gray clouds. With the play of *chiaroscuro* (light and shade) and the depiction of a half-drowned earth glazed in translucent brown tones, this illustration undoubtedly represented a cultural memory of inundation. Acknowledging water and its pervasive presence as crucial for the creation of a new biota, these narratives mythologically clime flood as "the Great Deluge" (Kalita 2013, 70). Be it the annual display of flooding or ceaseless rain, this clime's participation in monsoonal enactments received much creative attention within a rich literary tradition.

Until now, monsoon has been climed as a timely occurrence, regularly staging its drama, that formed a distinct more-than-monsoon history in this eastern Himalayan flatlands. A popular folk custom of holding frog weddings (*bhekuli biya*) to mitigate the rainfall crisis was one aspect of the monsoonal drama—one in which humans recognized frogs as ambassadors. One folk poem that comprises a dialogue between a female gardener (*maalini*) and a flower features rain and the croaking of frogs (*O megh o megh boroxun diyo kiya*). Likewise, the *Hudumdeu Puja* ritual was conducted during periods of prolonged drought. Naked women danced and chanted at midnight, plowing the barren fields to "tease" and disparage the Hindu rain god Indra (Sanyal 1965, 144; Datta 2012, 721). Indra was believed to be responsible for either sending or withholding rainfall. In Hindu theology, he and other rain gods are also associated with virility, pouring their fertile seeds onto the earth (Yoshinori and Shinde 2004, 287–290). Clearly, then, frogs, weddings, women, and fertility played important supporting roles that were integrated into the monsoonal drama for the humans of this clime. The Moran tribe, predominantly inhabiting the eastern part of the Brahmaputra Valley, contribute the traditions of stone wedding songs (*xeel biyar kheri*), songs of rain (*boraxunar kheri*), songs anticipating sunny days (*rodar kheri*), and songs of storms (*dhumuhar kheri*) to the drama (Rabha Hakacham 2017, 449–452). The following verse is taken from a song of storms:

> O *chuk bhekuli*, the clouds are here.[6]
> Call for the rain and give water to all.
> Look how it rains, like the *ou* flowers.[7]
> Joyous! All begin preparing *pithaguri.*[8]

Performances of these songs punctuated periods of drought—just one example of people translating their understanding and predictions of the weather into creative works. Numerous historical narratives record irregular rainfall or, occasionally, a complete lack of it, while the deep water-storage ditches in the Ahom royal palace are testament to the absence of rain showers at certain times of the year (Vansittart 1785, 471). Moreover, nearly every pre-modern kingdom across this clime used ponds and reservoirs to proclaim their sovereignty. *Buranji*s are full of anecdotes of digging "victory ponds" (*pukhuri*) as assertions of political power and to mark territorial boundaries. One seventeenth-century royal chronicle hints at a prolonged drought, although it fails to mention if this had fatal consequences for the population.[9] By contrast, earlier hagiographic records contain detailed accounts of the misery caused by a six-month long famine in the sixteenth century (Neog 1987, 194–195). Similarly, unbearable heat and perhaps a long hiatus between monsoonal dramas forced one mid-seventeenth-century Ahom king to change his attire, as well as his pillowcases, to white cotton (Bhuyan 1930, 45).

Popular aphorisms also hint at occasional droughts, which is helpful for climing monsoonal variables. Some of these advise farmers to sell their plows and seek employment in the royal court under certain circumstances. Others predict the weather: for example, if there is thunder in the north and a rainbow in the west, rain will fall in the east; if there is a rainbow in the east and thunder in the south, there will be rain in the west (Gogoi 1991, 30–31). A sixteenth-century popular poem sung in the central part of this monsoon clime tells of the excavation of a pond for a cuckoo in response to water scarcity (Saharia 2021, 24).

Therefore, while the majority of historical records do not explicitly mention any significant instances of drought, socio-cultural and ecologically determined customs present a different story. Whether through poetry, songs, or illustrations, humans and their literary expressions, both written and oral, climed the monsoon, perceived its staged drama, and cultivated their ecological comprehension to gauge the unpredictable rain, regardless of the efficacy of their methods.

Conclusion

This chapter has traced Assamese cultural memories of the monsoon, and its abundance or absence, portrayed as a "drama." It has explored the ways in which a watery clime was perceived visually, poetically, and historically in various pre-modern artistic productions, and has suggested a story of three monsoonal phases (from pre- to post-monsoon) that co-designed common monsoonal thriving across the eastern Himalayan foothills.

Although the monsoon was presented as a timely event in the official historical sources, various eco-cultural practices staged paradoxical dramas too. Irrespective of whether the monsoon arrived on time or not, people commonly discussed it through varied socio-cultural expressions. Recent scientific constructions (paleoclimatic studies) unfold the climatic variations affecting the staged monsoon drama. Accordingly, humans and nonhumans of the clime witnessed and internalized the "drama" with its timely or untimely showers, whether as audience or participant. From military strategy to experimental cropping patterns, monsoonal rains led the creation of this clime. With its annual rainfall and regular inundation, the Brahmaputra and its tributaries proved to be significant supporting actors in the building of a fertile rice zone. One major reason why the clime did not witness major crop failures was the timely monsoon and the regular trade along the river and its tributaries. Cultural portrayals of human–monsoon entanglements reflected an innate sense of Deep Time and history. With the monsoon pattern climed as a staged drama, and the monsoon as protagonist, climatic variables played a complex role in shaping a more-than-monsoon history across the eastern Himalayan foothills.

The chapter has enumerated a series of dramatic roles played by the monsoon, from the battlefield to bringer of deluge and drought. The history of human and other-than-human engagements in Assam's pre-modern literary productions offers myriad embodied, material, and imaginative meanings.

Notes

1 The Ahom–Mughal wars intensified from the first half of the seventeenth century and ended in 1682 CE. There are extensive records of them in Ahom royal chronicles (*buranji*s), including *Assam Buranji* by Harakanta Sadar Amin, *Tungkhungiya Buranji, Satsari Buranji, Deodhai Asam Buranji*, and so on.

2 *Padmapani Devar Charita*, JOR/MAJ/AUTIDPDG/40/198, 52a and b, SriSri Auniati Satra, Majuli.

3 Bihu, a seasonal folk practice, is celebrated three times each year—*Bohag* bihu in the spring, *Kaati* bihu in the autumn, and *Maagh* bihu in the winter.

4 The regional idioms are usually seen in figures in frontal view. They include three-quarter-profile faces, broad, fish-like eyes stretched toward the ears, and flat backgrounds with ornamented, multi-layered hills. The male and female forms are distinguishable only by style of dress.

5 *Sirish* (*Albizzia lebbek), karabi* (*Thevetia Peruviana), kanchan* (*Bauhinia Vahlii), keteki* (*Pandanus odoratissimus), kadam* (*Neolamarckia cadamba*), and so on.

6 *Chuk bhekuli* refers to the Asian common toad (*Duttaphrynus melanostictus*), which breeds during the monsoon.

7 *Ou* flowers are the white flowers of the elephant apple tree (*Dillenia indica*).

8 *Pithaguri* is ground rice flour, which serves as the base material for snacks prepared by every agricultural community of Assam.

9 *Deodhai Asam Buranji*, MS 42, 93a, Department of Historical and Antiquarian Studies, Guwahati.

References

Anonymous. n.d. *Chitra Bhagawata*. Guwahati: Srimanta Sankardev Kalakshetra. Digitised by Shri Krishna Museum, Kurukshetra.

Arnold, David and Ramchandra Guha, eds. 1995. *Nature, Culture, Imperialism: Essays on the Environmental History*. Delhi: Oxford University Press.

Barpujari, Heramba Kumar, ed. 1994. *The Comprehensive History of Assam*, Volume 3. Guwahati: Publication Board of Assam.

Behringer, Wolfgang. 1999. "Climatic Change and Witch-Hunting: The Impact of the Little Ice Age and on Metalities." *Climatic Change* 43(1): 335–351.

Bezbaroa, Lakshminath. 1988. "Burhi Air Sadhu" (Stories from Grandmother). In *Bezbaroa Granthavalee*, Volume 1, edited by Atulchandra Hazarika, 855–908. Guwahati: Sahitya Prakash.

Bhuyan, Suryya Kumar, ed. 1930. *Assam Buranji by Harakanta Sadar Amin*. Guwahati: Department of Historical and Antiquarian Studies.

Bhuyan, Suryya Kumar, ed. 1933. *Tungkhungiya Buranji*. Guwahati: Department of Historical and Antiquarian Studies.

Bhuyan, Suryya Kumar, ed. 1960. *Satsari Asam Buranji: A Collection of Seven Old Assamese Buranjis or Chronicles with Synopses in Assamese*. Guwahati: Publication Department, Gauhati University.

Blochmann, Heinrich. 1872. "Koch Bihar, Koch Hajo, and Asam, in the 16th and 17th Centuries, According to the *Akbarnamah*, the *Padshahnamah*, and the *Fathiyah-i-Ibriyah*." *Journal of the Asiatic Society of Bengal* 41: 49–101.

Bose, Shibani. 2020. *Mega Mammals in Ancient India: Rhinos, Tigers, and Elephants*. New Delhi: Oxford University Press.

Brandshaug, Malene K. 2023. "Water Climing: A Cosmopolitical Ecology of Water in the Southern Peruvian Andes." In *Storying Multipolar Climes of the Himalaya, Andes, and Arctic: Anthropogenic Climate and Shapeshifting Watery Worlds*, edited by Dan Smyer Yü and Jelle J.P. Wouters, 105–120. Abingdon and New York: Routledge.

Bristow, Tom and Thomas H. Ford. 2016. *A Cultural History of Climate Change*. Abingdon and New York: Routledge.

Carey, Mark and Philip Garone. 2014. "Forum Introduction." *Environmental History* 19(2): 282–293.

Chevalier, Jean-Baptiste. 2020. *Adventures of Jean-Baptiste Chevalier in Eastern India (1752–1765): Historical Memoir and Journal of Travels in Assam, Bengal and Tibet*, translated by Caroline Dutta-Baruah and Jeane Doluche. Guwahati and Paris: Lawyer's Book Stall Publications in association with EFEO.

Coleman, J.M. 1969. "Brahmaputra River: Channel Processes and Sedimentation." *Sedimentary Geology* 3(2–3): 129–239.

Cook, Edward R., Kevin J. Anchukaitis, Brendan M. Buckley, Rosanne D. D'Arrigo, Gordon C. Jacoby, and William E. Wright. 2010. "Asian Monsoon Failure and Meghadraught during the Last Millennium." *Science* 328(5977): 486–489.

Datta, Birendranath, ed. 2012. *Works of Prafulladatta Goswami*, Volume 2. Guwahati: Publication Board of Assam.

Dutta Baruah, Harinarayan, ed. 1950. *Chitra Bhagavata*. Nalbari: Dutta Baruah Brothers & Co.

Dutta Baruah, Harinaryan, ed. 1953. *SriSankar Vykyamrita*. Shillong: Assam Sahitya Sanmilan.

Gait, Edward. 1906. *History of Assam*. Guwahati: Lawyer's Book Stall.

Gogoi, Lila. 1991. *Asamiya Loka Sahityar Ruprekha: An Outline History of Assamese Folk Literature*. Dibrugarh: Students' Emporium.

Guha, Amalendu. 1982. "The Medieval Economy of Assam." In *The Cambridge Economic History of India*, Volume 1, edited by Tapan Raychaudhuri and Irfan Habib, 478–505. Cambridge: Cambridge University Press.

Guha, Amalendu. 1991. *Medieval and Early Colonial Assam: Society, Polity, Economy*. Calcutta: KP Bagchi & Company.

Gupta, Anil K., Som Dutt, Hai Cheng, and Raj K. Singh. 2019. "Abrupt Changes in Indian Summer Monsoon Strength during the Last ~900 Years and Their Linkages to Socio-Economic Conditions in the Indian Subcontinent." *Palaeogeography, Palaeoclimatology, Palaeoecology* 536. www.sciencedirect.com/science/article/abs/pii/S0031018219302974.

Hamilton, Francis. 1940. *An Account of Assam*. Guwahati: Department of Historical and Antiquarian Studies.

Hazarika, Manjil. 2017. *Prehistory and Archaeology of Northeast India: Multidisciplinary Investigation in an Archaeological Terra Incognita*. New Delhi: Oxford University Press.

Hulme, Mike. 2017. *Weathered: Cultures of Climate*. London: Sage Publications.

Irvine, Richard D.G. 2020. *An Anthropology of Deep Time: Geological Temporality and Social Life*. Cambridge: Cambridge University Press.

Kalita, Naren, ed. 2013. *An Illustrated Anadi Patana of Kuji-Satra (18th Century AD)*. Guwahati: Directorate of Museums.

Ladurie, Emmanuel Le Roy. 1971. *Times of Feast, Times of Famine: A History of Climate since the Year 1000*. New York: Doubleday.

Ludden, David. 2005."Where is Assam?" *Himal Southasian*, November 15. www.himalmag.com/where-is-assam/.

McCosh, John. 1837. *Topography of Assam*. Calcutta: G.H. Huttmann, Bengal Military Orphan Press.

McNeill, J.R. 2003. "Observations on the Nature and Culture of Environmental History." *History and Theory* 42(4): 5–43.

Mehrotra, Nivedita, Santosh K. Shah and Amalava Bhattacharyya. 2014. "Review of Palaeoclimate Records from Northeast India Based on Pollen Proxy Data of Late Pleistocene Holocene." *Quaternary International* 325: 41–54.

Nath, Rajmohan, ed. 1998. *Sankardeva-Madhavdeva Charit by Daityari Thakur*. Guwahati: Lawyer's Book Stall.

Nathan, Mirza. 1936. *Baharistan-i-Ghaibi*, Volume 2, translated by Maidul Islam Bora. Guwahati: Department of Historical and Antiquarian Studies.

Neog, Maheswar, ed. 1950. *SriSri Vamsigopaldevar Charitra by Ramananda Dwij*. Guwahati: Sri Sisir Sri Sankaradeva Library.

Neog, Maheswar, ed. 1987. *Guru Charita Katha*. Guwahati: Gauhati University Publication Board.

O'Hare, Greg. 1997. "The Indian Monsoon Part 1: The Wind System." *Geography* 82 (3): 218–230.

Paerregaard, Karsten. 2023. "Climing the Andes: Vertical Complementarity, Trans-human Reciprocity, and Climate Change in the Peruvian Highlands." In *Storying Multipolar Climes of the Himalaya, Andes, and Arctic: Anthropogenic Climate and Shapeshifting Watery Worlds*, edited by Dan Smyer Yü and Jelle J.P. Wouters, 53–68. Abingdon and New York: Routledge.

Parker, Geoffrey. 2013. *Global Crisis: War, Climate Change, and Catastrophe in the Seventeenth Century*. New Haven and London: Yale University Press.

Peal, S.E. 1882. "Note on Platform-Dwellings in Assam." *Journal of the Anthropological Institute of Great Britain and Ireland* 11: 53–56.

Pollock, Sheldon. 2006. *The Language of the Gods in the World of Men: Sanskrit, Culture, and Power in Pre-modern India*. Berkeley and London: University of California Press.

Pradhan, Rohit, Nimisha Singh, and Raghavendra P. Singh. 2019. "Onset of Summer Monsoon in Northeast India is Preceded by Enhanced Transpiration." *Scientific Reports* 9. www.researchgate.net/publication/337831800_Onset_of_Summer_Monsoon_in_Northeast_India_is_preceded_by_Enhanced_Transpiration.

Rabha Hakacham, Upen, ed. 2017. *Henā-Huchā Janajatīya Asmīyā Loka Sahitya*, Volume 2. Guwahati: Publication Board of Assam.

Rookmaaker, L.C. 1999. "Records of the Rhinoceros in Northern India." *Saugetierkundliche Mitteilungen* 44(2): 51–78.

Saharia, Kanak Chandra, ed. 2021. *Siyālā Gosāiñr Baṅsāvalī by Suchandai Kaviraj Misra*. Guwahati: Department of Assamese, Gauhati University.

Saikia, Arupjyoti. 2019. *The Unquiet River: A Biography of the Brahmaputra*. New Delhi: Oxford University Press.

Sanyal, Charu Chandra. 1965. *Rajbangsis of North Bengal*. Kolkata: Asiatic Society.

Vansittart, Henry. 1785. "A Description of the Kingdom of Assam (Taken from the *Alemgeernameth* of Mohammed Cazim)." *Asiatik Miscellany* 1: 459–482.

Vaudeville, Charlotte. 1986. *Bārahmāsā in Indian literatures: Songs of the Twelve Months in Indo-Aryan Literatures*. Delhi: Motilal Banarsidass.

Yasuda, Yoshinori and Vasant Shinde, eds. 2004. *Monsoon and Civilization*. New Delhi: Lustre Press and Roli Books.

7 A Thirsty Himalayas

Rain Clime and Anthropogenic Drought in the Darjeeling Hills

Sangay Tamang

Introduction

Himalayan clime studies, as laid out succinctly in Chapter 1 of this book, offers fresh approaches to global climate change by interconnecting it with local climate history, patterns, and evolution. It calls for multidimensional understandings of climate change and the ecological transcendence of the simplistic local–global dichotomy. Clime entails resituating humans as ecological and geological beings, apart from cultural beings, and foregrounds human terrestrial experiences to complement the abstract models of climate sciences that are largely divorced from diverse lifeworlds (Smyer Yü and Wouters 2023).

If clime is a place embodied with climate history, patterns, and change (Smyer Yü 2023, 8), I propose that the Darjeeling Hills region is best understood as a rain clime. Rain, an atmospheric phenomenon, is an inherent part of the earth's water cycle that is experienced terrestrially by humans and other-than-humans. My coinage of "rain clime" affirms rain as an agent of place-making. This is especially true in the Darjeeling Hills, a Himalayan foreland that receives extensive precipitation annually. Through rain clime as an embodied entanglement of rain and living beings, this chapter intends to demonstrate the history of Darjeeling as a Himalayan history of rain in colonial, botanical, political, economic, cultural, and spiritual terms. It presents rain as an agential power of the earth co-making indigenous knowledge and traditions, and as a modern, human-altered natural force that attests to the anthropogenic transformation of the Darjeeling landscape from colonial to postcolonial times. From a land deeply enmeshed in water, this anthropogenic transformation has made the hills "thirsty" for the first time. In what follows, I shall trace this transformation by focusing on "indigenous rain clime," "colonial rain clime," and "anthropogenic rain clime," each of which produces a distinctive narrative of rain, climate, place, and people in the Darjeeling Himalayas.

This chapter thus argues that Darjeeling's history—from the pre-colonial era through the colonial period to the postcolonial present—has been shaped by the entanglement and enmeshment of rain, humans, and nonhumans. I intend to elaborate this argument with a focus on the agential entanglement of rain with the multispecies worlds in the eastern Himalayas.

DOI: 10.4324/9781003484394-7

The Darjeeling Rain Clime

Himalayan mountain-making, as the earth's tectonic process, was also a process of climate-making (see Chapter 1, this volume). This is especially true in relation to precipitation. In broad terms, and when tracked through Deep Time, precipitation patterns pre- and post-Himalayan orogeny were fundamentally different. Historically, the formation of the Himalayas co-constituted a range of precipitation patterns. The western and eastern Himalayas bear two contrasting patterns of precipitation, often creating different forms of rainfall. Amrith and Smyer Yü (2023, 32) traced the historical role of uneven precipitation across the eastern, central, and western Himalayas in shaping the characteristics of monsoonal rain, and explained why the eastern Himalayas received more rain than snow compared to the western Himalayas, where larger volumes of water are stored in snow and glaciers. While highlighting the significance of the eastern Himalayas as a monsoon-maker, historians Pachuau and van Schendel (2022, 25) argue that

> the intense seasonality of the South Asia monsoon—very wet and hot from June to October, and mostly dry and cooler from October to June—has dominated the pace of life ever since. And the Eastern Himalayan Triangle stood out because it acted as a funnel, trapping more rain than other parts of South Asia.

The Himalayas have given birth to different kinds of rain climes (Iqbal 2023; Chapter 6, this volume), of which the rain clime of the easternmost Himalayan range occupies a central position in this chapter. The eastern Himalayan mountain ranges are important makers of rain during the monsoon. Thus, while the Himalayas, as rain-makers, provide a liquid lifeline to the alluvial plains in India, the monsoon also annually inundates the so-called Himalayan foreland of which Darjeeling is a part (on rain climes in the Naga hills, southeast of Darjeeling, see Chapter 2, this volume).

The Darjeeling Hills, located between other ranges of hills—in Nepal to the west, Bhutan to the east, and Sikkim to the north—are not merely recipients of rain but active rain-makers with a role in the making of the greater monsoon as well as local precipitation patterns and agency. The eastern Himalayas experience more rainy days than South Asia, ranging between 162 and 132 per annum in the Bhutanese Himalayas and the Darjeeling Himalayas, respectively (Prokop and Walanus 2017, 395). The longer monsoon, along with pre-monsoon and post-monsoon rainfall, keeps the region naturally cooler and also maintains the moist condition of the soil. This combination provides favorable condition for products like tea, cinchona, and cardamom. These cash crops came to define Darjeeling's engagement with and anthropogenic transformation by colonial and postcolonial states.

The Darjeeling Hills receive heavy rainfall mostly from June to September. This long monsoon season feeds the major rivers of the region, such as the

Teesta, Rangeet, and Balason, as well as many smaller rivers, streams, and springs. All of these waterways nourish numerous lifeforms. As the rainwater forms different kinds of climes, the physical environments of the Darjeeling Hills have evolved with the annual monsoonal rains. Furthermore, in addition to its environmental interactions with the physical earth, rain materially and spiritually nourishes different cultures in Darjeeling and the Sikkimese Himalayas (Bhutia 2023). Therefore, rain, an important agent of history, co-shapes physical landscapes, human knowledge of the environment, subsistence farming, transregional migration, and modern commercial activities that have anthropogenic implications for the Himalayan environment.

Notably, the British conquest and rule of the Darjeeling Himalayas from the nineteenth to the mid-twentieth century introduced commercial agriculture and its inevitable anthropogenic impact to the region. The British colonization of the eastern Himalayas and Bengal could be construed as a process of colonizing the rain climes of both the highlands and the lowlands, interlinked by the monsoonal rains and the transregional rivers. The British implemented modern scientific methods to harness, manage, and channel water resources as per the need of their capitalist exploitation of South Asia. This scientific knowledge gradually displaced the "indigenous rain clime"—an inherently living dynamic—both as an earth force and as an "anthropogenic rain clime" where rain was stored, channeled, and distributed through a network of pipes. Perhaps counter-intuitively, this process led to water scarcity.

While rain is a life-giver in the hills, it is also a destructive force in the context of modern, human-induced climate change. Since the colonial transformation of the landscape through various commercial activities, the Darjeeling Hills have witnessed countless landslides, and the rain clime has thus consisted of shifting terrains through floods and landslides. The rest of this chapter will explain how I understand the rain clime of the hills in indigenous, colonial, and anthropogenic terms through historical and ethnographic narratives. Fieldwork was conducted between March and June 2023 in the Darjeeling Hills in order to understand the entanglement of rain and the water crisis. However, the chapter also draws on extensive ethnographic fieldwork conducted between 2017 and 2020 as part of the author's doctoral thesis.

The Indigenous Rain Clime

If clime is seen as a place, and if place is seen as clime, a rain clime is a situated between rain, place, and climate changes, with each implicating the others. Thus, the Darjeeling Himalayas, as a clime, are not just passive recipients of rain but also makers and movers of rain. By "indigenous rain clime," I mean both the geologically ancient Himalayas and their native peoples. In the geological sense, an indigenous clime is an inherent part of what Smyer Yü (2023, 14) calls the "indigenous earth"—or the Deep Time earth—which is significantly older than the current earth that humans have dominated over the last 200,000 years. In the human sense, "indigenous rain clime" refers to the Darjeeling Hills as a culturally ancient living environment of several Himalayan indigenous human societies. This is particularly important because the indigenous human inhabitants hold diverse knowledges of the Himalayan climate in material, spiritual, and cultural terms. These two conceptions of "indigenous rain clime" have been interwoven with one another. At the same time, they have been anthropogenically transformed by colonial conquest and postcolonial economic development. In many ways, the Darjeeling Hills are suffering from what I call "anthropogenic thirst," which I elaborate in the final section of the chapter.

In this context, the indigenous rain clime is not only pre-colonial, where the rain is not influenced by modern anthropogenic factors, but affirms rain as both a generative and a destructive force, and locates its centrality to indigenous knowledge, folklore, oral tradition, and ritual. Knowing and predicting rainfall patterns was central to human survival in the Darjeeling Hills and the surrounding mountains. This centrality is reflected in indigenous wisdom and rituals. For instance, the Lepcha, an indigenous community of Sikkim and Darjeeling, have a rich tradition of worshipping rain in a festival called Tendong Lho Rumfaat—the "Prayer of Tendong Mountain"—during which rain is portrayed as an important agent of eco-cultural change in the Himalayas. The Lepcha commemorate their ancestors who climbed a mountain called Tendong (Uplifted Horn)—which is now worshipped as a god—in order to survive 40 days and 40 nights of continuous rain that submerged all the other peaks in the region. In the Lepcha's mythological world, water is the ecological essence of the living world of plants, animals, and humans. It is celebrated in various religious and cultural performances as the life-giver of everything. For example, the community's ancestors are said to have been made from snow on the summit of Mount Kanchenjunga, which is the source of the Teesta and Rangeet rivers. Water spirits, in the form of animals or birds, shaped the movements and courses of these rivers.

Rain is articulated and imagined in various forms. In the form of precipitation, it is culturally and spiritually experienced and conceptualized throughout the region, with mist construed as the foothills worshipping the mountains (Bhutia 2023). Also, rain in the form of bodies of water such as lakes, springs, and rivers serves different spiritual and healing practices. For instance, among the Lepcha of Dzongu Sikkim, it is believed that, "in case of a cut or a wound, wash in a stream and the wound would be healed" (Lepcha 2021, 47). When rain falls on the ground and flows through natural channels, it is considered pure or holy. Many Nepali consider water flowing from natural sources as *chokho pani* to contain healing properties and spiritual essence. It is also used to heal those "who become suddenly frightened as a result of some known or unknown incident" (Lama and Rai 2016, 93). During my fieldwork in the Darjeeling Hills, one repsondent told me:

> Once, I was suffering from chronic allergies. I visited many doctors and took lots of medicine, but it didn't heal completely. If I stopped taking

> the medicine, [the rash] would reappear on my skin. After years of suffering, my neighbor suggested I should visit a local shaman [*jhakri*]. Upon seeing my allergy, he told me to offer milk or *chokho pani* to Nag Devta [a *naga* deity] every Monday. I immediately started visiting the spring and offering milk in the area where the water originated. Miraculously, my allergy completely healed thereafter.

The ground, which stores rainwater and facilitates its flow through natural channels (mostly springs, rivers, and streams), is a sacred space in the Darjeeling Himalayas, where different religious groups maintain their differential but common relationship with water. Indeed, it is regarded as a space where various deities reside, so it may be considered as a site of *sarvadharma* (literally meaning "all religions coexist").

This perception of rain as a gift from the earth has shaped the local communities' understanding of the regional ecology and the natural climate cycle. Akin to the Lepcha, communities of Nepali origin in the eastern Himalayas enact their deep connection with rain through rituals like Sansari Puja. During this festival, which is usually held in Baishak (the first month of the Nepali calendar), shamans known as *jhakri* or *bahun* petition Sansari Mata—the mother of all creation—for timely rain and an abundant harvest. Therefore, the ritual is based on three assumptions: that nature is a divine creation; that the human body is intimately connected to it; and that humans must ask the gods for timely and adequate rain. This making and mediating of rain by ritual specialists through the performance of rituals like Sansari Puja highlights the commonality of rain rites through which diverse religious and ethnic groups solicit rain and an abundance of water. Thus, the indigenous rain clime affirms the sacred, as most rain-storage and channeling places, such as springs (*dhara*), lakes, and rivers, are considered sacrosanct, while the water they contain is viewed as pure and holy (*choko pani*).

Similarly, other festivals, such as Asar Pandra and Saune Sankranti, not only reflect rain's intimate relationships with humans, plants, animals, and other natural forces but serve as important drivers of local knowledge of climate and weather. In many areas of the eastern Himalayas, these ways of perceiving rain through rituals or festivals indicate the arrival of the monsoon, the season of rice planting, and, most importantly, the local understanding of climate. In this indigenous rain clime, changes in climate are also perceived through cosmic knowledge. Late rain or a clear sky during the monsoon symbolizes inauspiciousness, as expressed in aphorisms like, "Sunny days in the months of June, July, and August are not auspicious." Also, most natural springs (*dhara*)—primary sources of water in the hills—depend on rainfall during the monsoon; therefore, less rain at that time of year means serious thirst across the hills. Many locals express their concern by saying, "If there is not enough rain during monsoon, what will we drink throughout the year?" The monsoon, as received, imagined, and performed through various communities' rites, offerings, and rituals, plays a vital role in shaping the flow of water in the Himalayas.

While the arrival of the monsoon is a life-generating force in the Himalayas, it also brings unforeseen disaster in the form of landslides. Indra Bahadur Rai described the impact of one such landslide in "Long Night of Storm":

> Now, recalling some strength, Kaley's mother said, "Peril and calamity are everywhere. Yes, the storm ruined things but we will fix everything now—it is nothing impossible to do. We have our own home, a cattle-shed full of cows, a field, and in the field some thirty or forty groves of bamboo, gooseberry, and fig trees, and cucumber vines reaching the skies. How much more ruin can a storm bring? See, I have to rush now, get home and rebuild."
>
> (Rai 2018, 147)

This local perception of rain is a result of sustained environmental degradation since the British colonization of the hills. In their pursuit of turning Darjeeling into a colonial hill station, the British appropriated rainwater through a variety of scientific measures, which contributed to the creation of the "colonial rain clime" and ultimately the "anthropogenic rain clime" of the modern eastern Himalayas.

Rain and the Making of Colonial Darjeeling

In this section, I present rain as an agent in the making of colonial Darjeeling. While the incessant monsoon created a tremendous obstacle for the colonial state's assent into Himalayan regions like the Darjeeling Hills, it was also rain and the cooler climate of the Himalayas that led the British into the hills. The scientific measurement of rain, analysis of the soil, taxonomic categorization of plants and animals, and the recording of landslides and other natural disasters were ultimately carried out by the British, who wished to catalog and secure Darjeeling as a profitable and habitable colonial station that also fulfilled their need for a safe, upland-England-like (in terms of climate, that is) "sanatorium" to heal European convalescents.

It was ultimately rain and its climatic cooling effects that drew in the colonial state and gave rise to what we may call "the medicalization of rain." British officers experienced the scorching Indian plains as dirty, crowded, and replete with diseases, including malaria, tuberculosis, and cholera. Besides medicine, the prescribed treatment generally included a period of time in a hill station, of which Darjeeling became one. The European fear of the Indian climate was legitimate: "between 1859 and 1867 a third of all deaths among British troops in India were due to disease" (Arnold 2004, 127), and these diseases were readily linked to India's "unhealthy" climate. Therefore, the Darjeeling Hills, with their climatic contrast to the plains, appeared to be an ideal location for the British to base their summer capital (Bhattacharya 2012).

After the Anglo-Gurkha War (1814–1816), from which the British emerged victorious, surgeons and other medical officers and staff were sent to Darjeeling to evaluate the region's suitability as an open-air British hospital and sanatorium. In 1828, two colonial officers, Captain Lloyd and Lieutenant General George W. Aylmer Lloyd, judged the climate to be favorable to the healing of European bodies. This was part of the imperial rationale behind the East India Company's decision to seize Darjeeling from the Rajah of Sikkim in 1835 (O'Malley 1985). In this making of Darjeeling as a clime of healing, the notions of climate, rain, and health took precedence. In his note on Darjeeling, the surgeon J.T. Calvert (1909) wrote:

> From the beginning of June till the middle of October there are heavy rains with mist and absolute saturation of the air with moisture. There is not much sunshine and exercise out of doors is curtailed. To keep out the mist and damp, rooms have to be shut up, and as there is not much air current, the house gets stuffy. During the period mentioned, I should say the climate was unsuitable for all classes of *medical* cases. From October to the middle of May I should regard the climate as very suitable for early cases.

Likewise, Dr. Captain J.D. Herbert, Deputy Surveyor General, wrote, "the first point to be considered in the establishment of a station of health is obviously climate" (Herbert 1830, 3). The Darjeeling Hills, therefore, became a refuge from the tropical weather and deadly diseases of the plains. As Sir James Ranald Martin (1796–1874) remarked: "[The mountains] are especially protective against the diseases of the plains; they are curative in simple fevers, unaccompanied by organic disease; visceral diseases are rarely cured in the mountains" (quoted in Dasgupta 2023, 54).

While imperial and military objectives have been generally highlighted in the making of colonial Darjeeling, I suggest that climate, and especially rain, was a prime agent in the annexation and subsequent administration of the area. Many historians have argued that the creation of hill stations for the well-being of European convalescents was part of the larger colonial economy (Bhattacharya 2012, 25). However, the significance of rain and its clime in accelerating the colonial process in the hills is seldom discussed.

Early studies on climate and rain by various colonial naturalists, ethnologists, and surgeons like Archibald Campbell, J.D. Hooker, Brian Hodgson, and many others supported the British decision to base their capital in the hills. Before Darjeeling, the British had attempted to establish a hill station at Cherrapunji (then Assam, now Meghalaya). However, its rain was considered to be too extreme—it is the wettest place on earth—whereas Darjeeling's precipitation pattern was deemed to be "just right" for a sanatorium (Pinn 1986) as well as, later, for tea plantations. The selection of Darjeeling as an ideal location for a hill station therefore came after the wide circulation of medical reports and naturalists' studies, as well as personal observations of the climate.

The British annexation of the Darjeeling area connoted the beginning of the colonial rain clime, in which precipitation and climate became primarily understood in relation to British health and revenue-making. The latter interest came into play in relation to the tea plant. The moist soil was suitable for growing crops like maize, wheat, barley, rice, and cardamom. However, all of these were considered unprofitable, so new, more commercial crops were sought to make the hills "productive" and "profitable." Dr. Campbell experimented with various moisture-loving crops in his backyard nursery at Beechwood, Darjeeling, and after repeated failures with cotton, he finally succeeded in cultivating Chinese tea plants (*Camellia sinensis*). This came at a time when the British were in need of alternative sources of tea because of political tension between the East India Company and China. Thus, the colonial government encouraged the crossing of Chinese tea plants with a wild Assamese variety in a number of nurseries in the hills. Therefore, the Darjeeling Hills became a natural laboratory in which the British experimented with botanical species, and on which they brought their scientific knowledge to bear.

The successful cultivation of tea in Darjeeling was enabled by the expansion of scientific knowledge on climate, species, and plantation agriculture. Natural resources, mostly forest, were subjected to extensive scientific investigation (Guha 2010), including measuring rainfall, to determine the survival and flourishing of various botanical species. For instance, European tea planters, as well as foresters, started observing and recording rainfall data and studied the variation of rainfall in the Darjeeling Hills. As elsewhere, these studies led "to a form of environmental determinism that excludes human factors" (Carey and Garone 2014, 290).

This climate science was part of the imperial project to make the Darjeeling Hills and mountains a profitable tea-producing area. The appropriation of rain for the commercialization of tea in the colonial rain clime became juxtaposed with the indigenous rain clime and the indigenous cultivation of traditional crops like cardamom that depended on local irrigation channels from streams. Thus, it was rain through the processes of its channeling, storage, and measurement that (ecologically) divided the hills into plantation and "outside plantation" (Besky 2021) zones. This bifurcation has had a significant impact on both human settlement and landscape formation since the colonial era.

The planting of tea required rapid deforestation and the introduction of plantation-style mono-cropping, thus changing the "look" and "feel" of the land. The focus on tea also meant the privileging of certain other species. For instance, the quick-growing Japanese cedars (*Cryptomeria japonica*) was planted on a large scale to meet the high demand for timber from the tea plantations' and the railways. Once again, this transformed the landscape of the Darjeeling Himalayas. Moreover, these trees are sometimes portrayed as an environmentally negative (Rai and Joachim 2018), not least because they "require a lot of water for gas exchange during the early spring" (Mori et al. 2022). The imperial privileging of this species thus impacted flows of water, as did deforestation, which resulted in increased run-off and more landslides. In his report on Kalimpong, C.A. Bell (2022, 62) mentioned that, "in 1902, the landslips were larger and more frequent than they had been for many years past." Clearly, these landslides were anthropogenic in nature. Similarly, some years earlier, the *Annual Administrative Reports of the Railways in India 1885–1886* had observed that "a continuous 45 inches of rainfall for ten days between 29 June and 8 July 1885 caused 20 landslips along the hill slopes, interrupting railway traffic in Darjeeling" (quoted in Dasgupta 2023, 166).

In addition to this natural and ecological transformation of the hills, the colonial rain clime came to influence the ethnic make-up of the human inhabitants through the importation of laborers—variously bonded and enslaved—who were transported to work on the plantations. In the colonial rain clime, climate and race congealed into a hierarchy of labor in which tribes from Chotanagpur, who were first recruited and worked on plantations in Assam, became seen as less suitable given the incessant monsoon and cool climate of the Darjeeling Hills, to which they struggled to habituate themselves. Many were forcibly rounded up only to run away at night (Lama 2008, 86; Tamang 2022, 8). This led the colonial state to look to the oppressed and marginalized Kiranti ethnic groups of eastern Nepal, who acclimatized more readily to the topography and cold weather of the Darjeeling Hills.

Thus, while it was rain that led to the colonial transformation of the hills into tea plantations, it was climate and racial prejudice that informed the labor recruitment that determined the tribal–ethnic demography of Darjeeling over the long term. The similarities between the two regions' topographical and climatic conditions enabled many ethnic groups of Nepali heritage from eastern Nepal to adjust quickly to the clime of the Darjeeling Hills. Thus, it was the rain clime that brought both tea plants and different ethnic laborers to Darjeeling—the experience and knowledge of which differed greatly between colonizer and colonized. This antecedently set the stage for later ethnic and political movements that have been discussed at length elsewhere (Subba 1992).

As many Nepalis labourers along with their predecessor and other indigenous groups started to settle in the region, their traditional knowledge led to an expansion of the indigenous rain clime. Additionally, these laborers, although of different ethnic backgrounds, came to share common ecological experiences. The cultivation of wet rice on the hilly slopes was popular among many Nepali laborers, and they not only brought their rain rites and festivals like Asar Pandra to Darjeeling but also established the domination of Nepali cultural and linguistic influence in the region. Perhaps surprisingly, this did not generate major ethnic tension, but rather led to the development of a commonality of indigenous rain climes with similar ecological and topographical experiences. For instance, O'Malley (1985, 64) wrote, "from Nepali they [the Lepcha] have also learnt how to construct on the mountain slope the terraces." Meanwhile, the Nepalis adopted many Lepcha agricultural techniques and started to use the likes of cardamom. Therefore, the banks and beds of Darjeeling's streams, where the abundant water supply enabled the cultivation of rice, cardamom, and maize, became preferred settlement locations for many Nepali tea-plantation laborers.

In all of this, it was rain in its varied forms that enabled the British to turn the Darjeeling Hills into an important site of imperialist and capitalist venture in the eastern Himalayas. However, the large-scale deforestation, monocropping, urban development, and migration (both European and local) that followed in the wake of the hills' transformation into a hill station and plantation zone caused a serious ecological crisis. These changes to the rain clime during the colonial era had a drastic effect on local precipitation patterns and contributed to the breakdown of traditional watersheds. Water's reduction to its commercial, political, and technological properties spurred anthropogenic transformations and ultimately contributed to the "thirst" of modern-day Darjeeling.

The Anthropogenic Thirst of the Darjeeling Himalayas

If the indigenous rain clime was characterized by an abundance of water to which cultural, ritual, and spiritual life was variously attuned, and if the colonial rain clime was predicated on appropriating that water for imperial use, the postcolonial story of contemporary Darjeeling is one of water shortages and water politics. It is often contended that water shortages are a product of political mismanagement, with planners insisting they are caused by a lack of funds, scientists attributing them to climate change, and politicians promising water in every household prior to each election. This section presents the anthropogenic rain clime as a byproduct of the colonial rain clime and a weakening of the indigenous rain clime that resonates multi-dimensional causes of water scarcity in the region. In the anthropogenic rain clime, in addition to being stored, appropriated, and managed through piped networks, water moves through a political and financial network that reinforces inequality, marginalization, and exclusion, thereby making residents subject to what Nikhil Anand (2017) calls "hydraulic citizenship," with Darjeeling acting as a "hydraulic city." The rapid acceleration of piped water connection in the region shifted the discourse of water toward this anthropogenic hydrology in which natural springs (*dhara*)—which hold great eco-spiritual and socio-cultural significance among many indigenous communities—became less relevant.

Furthermore, relentless urbanization, the consequent growth of the urban population (seasonal as well as permanent), and deforestation in postcolonial Darjeeling have resulted in a steady decline in the number of springs and their discharging capacity, which in turn has led to acute water shortages in the region (Chhetri and Tamang 2019, 242). Thirty-two major springs have been identified in and around Darjeeling town (Boer 2011), but many of them are in a state of destruction. In the absence of springs and other natural sources of water, the notion of spring water as pure—or *chokho pani*—is replaced by what I call here "anthropogenic water." Studies of the drinking-water crisis in Darjeeling have shown that it is closely associated with declining local water sources, mismanagement of the municipal water supply, booming water businesses, and so on (Lama and Rai 2016; Drew and Rai 2016; Tamang et al.

2020; Chhetri and Tamang 2019; Shah and Badiger 2018). Although the indigenous rain clime remains, the indigenous flows of water have been variously altered, fragmented, damaged, and ultimately replaced by anthropogenic water, with "anthropogenic" here implying stored, contained, and channeled through human engineering, political networks, piped connections, and other means. Anthropogenic water, then, is characterized by the drying up of indigenous water.

With focus shifting toward anthropogenic water systems, the piped network introduced during the colonial period not only expanded but became critical for the viability of life in Darjeeling in the postcolonial era. Between 1910 and 1930, two artificial lakes were constructed below the present-day Senchel Wildlife Sanctuary, 15 kilometers upstream from Darjeeling (Drew and Rai 2016, 324), into which spring water from the surrounding catchment area was collected for distribution to the town. This modern system of water supply through pipes was designed to cater to the needs of the town's (primarily European) 30,000 residents, who were housed in private residences, hotels, the sanatorium, and the garrison. However, it is wholly inadequate to meet the needs of the current population, who require an estimated 1,970,000 gallons (8,955,797 liters) per day (Darjeeling Municipality Waterworks Department 2014). Moreover, such water as there is not equally distributed among the town's residents. The exclusion of marginalized sections of the community from piped water (Shah 2022) evolved not just through complex hydraulic infrastructure, planning, and engineering embedded in the political nexus but also through the classification of formal and informal water sources.

The municipality's piped water system is defined as formal water connection, in contrast to traditional water sources (springs), water tankers, and water porters, which are considered informal. However, this classification is not rigid as it is full of entanglements, especially in recent years, which have witnessed an enormous effort to channel all kinds of water through piped networks as well as the commercialization and privatization of traditional sources. It has been argued that only about 15 percent of the residents of Darjeeling town rely on the formal water supply, with the rest sourcing their water from informal connections, such as springs, water tankers, water porters, and so on (Shah 2022, 634). Therefore, categorizing the water supply as formal and informal—with the former being anthropogenic and the latter indigenous—merely reinforces inequalities and limits the provision of administrative support to only a small proportion of water users. It might be more productive to rethink the water crisis in the region in terms of anthropogenic and non-anthropogenic water, a distinction that rests on the historical, cultural, material, and spiritual relations people have with monsoon, rain, and water.

Having said this, one might argue that the water crisis in the Darjeeling Hills is not merely the result of the failure of hydraulic infrastructure; it is also due to the push toward anthropogenic (piped) water, the changing landscape over the last century, and the urbanization and concretization of catchment areas (Banerjee 2023), all of which have weakened and threatened the indigenous rain clime. The inability of the existing infrastructure to meet the population's demand for water is well recognized by the civic authorities (see Darjeeling Municipality Waterworks Department 2014), but the way the piped water connection is perceived as symbolic of status by urban dwellers is reinforcing inequalities and marketization and aiding the adoption of anthropogenic water.

Most of the natural springs in and around Darjeeling are owned by the community, with very few in private hands. Maintenance of the water supply is the responsibility of community (or rather village) organizations called *samaj* (Lama and Rai 2016). Although these groups still conduct rain rites like Sansari Puja, keep the springs clean, and plant sacred species—all important aspects of the indigenous rain clime—they have undergone substantial structural and functional changes over recent years. For instance, many *samaj* have constructed cement tanks to store spring water for use in the dry season. However, there have been complaints that pipes have been laid to carry water from the tanks to the homes of elite members of the *samaj* (usually the president and secretary) and/or the owner of the land on which spring is located, but not to other households. Consequently, most people still have to fetch their water by hand. One member of a *samaj* reported:

> The president and secretary of the *samaj* have private water connections at their homes and they rarely need to visit the *dhara*. The only time they visit the *dhara* is when there are some rituals at their home for which *chokho pani* is required. If they do not need to use the community line [a public hydrant] to fetch drinking water, then why would they even bother to talk about the issue of water?

The postcolonial story of water in Darjeeling is one of the gradual disappearance of *dhara*, and of the use of spring water becoming "merely" ritual, as opposed to both ritual and everyday. Also, due to the massive outmigration of youth, who play a vital role in fetching water from springs, the pipe network has become critical for life in Darjeeling, whether in the form of the formal water supply or through informal sources. In both cases, the further development of the pipe network is affecting the natural flow of water, the local perception of rain, and the indigenous rain clime, and thus driving the region toward an unsustainable future.

This transition toward anthropogenic water sources, where rain is contained, collected, and commercialized, has accelerated shortages and drought, making the hills thirstier than ever before. Today, as is the case in many other metropolitan areas, Darjeeling's network of pipes is a key site through which social status is achieved, the legitimacy of local administrators and politicians is evaluated, and residents assert their connection with water and rain (see Anand 2017, 10). The augmentation of piped water, which comes with all sorts of anthropogenic impacts, has heavily shaped the flow of water in today's Darjeeling Hills. No doubt, climate change is causing delays in the monsoon, unpredicted snowfall, and changing precipitation patterns, concurrent with the rampant concretization, deforestation, and other ecological crises that have affected the flow of water in the catchment areas. This, along with the human population's increasing detachment from natural springs has resulted in severe anthropogenic thirst in the region. As a consequence, indigenous water is being replaced by anthropogenic water across the hills of Darjeeling.

The water crisis demands multilayered engagement with various aspects of ecological, eco-spiritual, political, financial, and hydraulic infrastructure. Hence, it needs to be understood and studied in terms of its interrelationship with the clime or "rain clime" approach to open a space to rethink climate change and its impact on water.

Conclusion

This chapter enlivens Darjeeling's rain clime by conceptually distinguishing between "indigenous," "colonial," and "postcolonial" rain climes. It innovates the notion of "rain clime" to critically capture the agential entanglements between rain, landscape, humans, and other-than-humans in the Darjeeling Hills. The story that unfolds is significantly one of indigenous water changing to anthropogenic water, from an indigenous rain clime (in both earth and human terms) to an anthropogenic rain clime mediated by the colonial rain clime that altered, appropriated, and channeled the flow of water through the history of Darjeeling. While the process of colonization is often presented as the conquest of land, territory, and people, in this chapter it is reframed as the conquest (in the sense of appropriation and management) of climate, and especially of precipitation. This has affected rain in terms of both its quality and quantity, and in the sense of where it meanders and is stored, channeled, diverted, and absorbed after falling to the ground.

While rain has historically been abundant—as indigenous storytelling, ritual, and praxis testify—the process of colonization has led to anthropogenic droughts in the present day. This, along with political, technological, and infrastructural development, has made the Darjeeling Hills thirsty hills, thereby propelling communities toward engineered water supplies that are emblematic of anthropogenic water. Although the push toward anthropogenic piped water perhaps may have improved the accessibility of water among those who previously had to carry it long distances over rugged terrain, completely ignoring the significance of natural sources has greater consequences in various aspects of hydraulic life in the Himalayas. As a result, contrary to its earlier watery abundance, the town of Darjeeling has today grown thirsty.

References

Amrith, Sunil and Dan Smyer Yü. 2023. "The Himalaya and Monsoon Asia: Anthropocenic Climes since the 1880s." In *Storying Multipolar Climes of the Himalaya, Andes and Arctic: Anthropocenic Climate and Shapeshifting Watery Lifeworlds*, edited by Dan Smyer Yü and Jelle J.P. Wouters, 29–51. Abingdon and New York: Routledge.

Anand, Nikhil. 2017. *Hydraulic City: Water and the Infrastructures of Citizenship in Mumbai*. Durham, NC: Duke University Press.

Arnold, Edwin. 2004. *India Revisited*. London: Adegi Graphics LLC. (First published 1886.)

Banerjee, Amitava. 2023. "Unplanned Urban Sprawl Takes Toll on Darj's Natural Water Tower." *Millenium Post Online*, June 11. www.millenniumpost.in/bengal/unplanned-urban-sprawl-takes-toll-on-darjs-natural-water-tower-521756.

Bell, C.A. 2022. *Final Report on the Survey and Settlement of Kalimpong Government Estate in the District of Darjeeling*. Siliguri: NL Publishers. (First published 1905.)

Besky, Sarah. 2021. "The Plantation's Outsides: The Work of Settlement in Kalimpong, India." *Comparative Studies in Society and History* 63(2): 433–463.

Bhattacharya, Nandini. 2012. *Contagion and Enclaves: Tropical Medicine in Colonial India*. Cambridge: Cambridge University Press.

Bhutia, Kalzang Dorjee. 2023. "Offerings from the Rivers to the Mountains: Mist and Fog as Connecting Life Force in the Sikkimese Himalayas." In *Storying Multipolar Climes of the Himalaya, Andes and Arctic: Anthropocenic Climate and Shapeshifting Watery Lifeworlds*, edited by Dan Smyer Yü and Jelle J.P. Wouters, 121–137. Abingdon and New York: Routledge.

Boer, L. 2011. *Perennial Springs of Darjeeling: A Survey to Community Based Conservation*. Darjeeling: Ashoka Trust for Research in Ecology and the Environment.

Calvert, J.T. 1909. "Note on Darjeeling Climate in the Treatment of Phthisis." www.ncbi.nlm.nih.gov/pmc/articles/PMC5167875/pdf/indmedgaz71568-0033.pdf.

Carey, Mark and Philip Garone. 2014. "Forum Introduction." *Environmental History* 19(2): 282–293.

Chhetri, Ashish and Lakpa Tamang. 2019. "Decentralization of Water Resource Management: Issues and Perspectives Involving Private and Community Initiatives in Darjeeling Town, West Bengal." *Annals of the National Association of Geographers India* 39(2): 240–255.

Darjeeling Municipality Waterworks Department. 2014. *A Report on Water Supply System of Darjeeling Municipal Area*. Darjeeling: Darjeeling Municipality Waterworks Department.

Dasgupta, Dipanwita. 2023. *The Ravaged Paradise: Environmental History of Colonial Darjeeling Himalayas (1835–1947)*. New Delhi: Manohar Publishers and Distributors.

Drew, Georgina and Roshan P. Rai. 2016. "Water Management in Post-Colonial Darjeeling: The Promise and Limits of Decentralised Resource Provision." *Asian Studies Review* 40(3): 321–339.

Guha, Ramachandra. 2010. *The Unquiet Woods: Ecological Change and Peasant Resistance in the Himalayas*. Ranikhet: Permanent Black. (First published 1989.)

Herbert, J.D. 1830. *Report on Dargeeling: A Place in the Sikkim Mountains, Proposed as a Sanitarium, or Station of Health*. Calcutta: Baptist Mission Press.

Iqbal, Iftekhar. 2023. "Life and Loss of a Felt Habitat: Exploring the World of *Haor* in Bangladesh." In *Storying Multipolar Climes of the Himalaya, Andes and Arctic: Anthropocenic Climate and Shapeshifting Watery Lifeworlds*, edited by Dan Smyer Yü and Jelle J.P. Wouters, 138–154. Abingdon and New York: Routledge.

Lama, B.B. 2008. *The Stories of Darjeeling*. Kurseong: Nilima Yonzone Lama Publications.

Lama, Mahendra P. and Roshan P. Rai. 2016. "*Chokho Pani*: An Interface between Religion and Environment in Darjeeling." *Himalaya* 36(2): 90–98.

Lepcha, Charisma K. 2021. "Lepcha Water View and Climate Change in Sikkim Himalaya." In *Environmental Humanities in the New Himalayas: Symbiotic Indigeneity, Commoning, Sustainability*, edited by Dan Smyer Yü and Erik de Maaker, 43–65. Abingdon and New York: Routledge.

Mori, Hideki, Kana Yamashita, Saiki Shin-Taro, Asako Matsumoto, and Tokuko Ujino-Ihara. 2020. "Climate Sensitivity of *Cryptomeria japonica* in Two Contrasting Environments: Perspectives from QTL Mapping." *PLoS One* 15(1): e0228278.

O'Malley, L.S.S. 1985. *Bengal District Gazetteers: Darjeeling*. New Delhi: Logos Press. (First published 1907.)

Pachuau, Joy L.K and Willem van Schendel. 2022. *Entangled Lives: Human–Animal–Plant Histories of the Eastern Himalayan Triangle.* New Delhi: Cambridge University Press.

Pinn, F.M.J. 1986. *The Road of Destiny: Darjeeling Letters 1839*. Calcutta: Oxford University Press.

Prokop, Pawel and Adam Walanus. 2017. "Impact of the Darjeeling–Bhutan Himalayan Front on Rainfall Hazard Pattern." *Natural Hazards* 89: 387–404.

Rai, Indra Bahadur. 2018. *Long Night of Storm*, translated by Prawin Adhikari. New Delhi: Speaking Tiger.

Rai, R.K. and Joachim Schmerbeck. 2018. "Why Forest Plantations Are Disputed? An Assessment of Locally Important Ecosystem Services from the Cryptomeria japonica Plantations in the Darjeeling Hills, India." In *Conifers*, edited by Ana Cristina Gonçalves, 113–126. London: Intech Open.

Shah, Rinan. 2022. "Marginality and Informality in Domestic Water Scarcity: Case of a Self-Service Mountain Town." *Asian Studies, The Twelfth International Convention of Asia Scholars* 1: 630–639.

Shah, Rinan and Shrinivas Badiger. 2018. "Conundrum or Paradox: Deconstructing the Spurious Case of Water Scarcity in the Himalayan Region through an Institutional Economics Narrative." *Water Policy* 22(S1): 146–161. Smyer Yü, Dan. 2023. "Multipolar Clime Studies of the Anthropocene Himalaya, Andes and Arctic: An Introduction." In *Storying Multipolar Climes of the Himalaya, Andes and Arctic: Anthropocenic Climate and Shapeshifting Watery Lifeworlds*, edited by Dan Smyer Yü and Jelle J.P. Wouters, 1–26. Abingdon and New York: Routledge.

Smyer Yü, Dan and Jelle J.P. Wouters, eds. 2023. *Storying Multipolar Climes of the Himalaya, Andes and Arctic: Anthropocenic Climate and Shapeshifting Watery Lifeworlds.* Abingdon and New York: Routledge.

Subba, T.B. 1992. *Ethnicity, State and Development: A Case Study of Gorkhaland Movement in Darjeeling*. New Delhi: Har-Anand Publications.

Tamang, Lakpa, Ashish Chhetri, and Abhijit Chhetri. 2020. "Sustaining Local Water Sources: The Need for Sustainable Water Management in the Hill Towns of the Eastern Himalayas." In *Water Management in South Asia*, edited by Sumana Bandyopadhyay, Sucharita Sen, Habibullah Magsi, and Tomaz Ponce Dentinho, 123–131. Cham: Springer.

Tamang, Sangay. 2022. "Environmentalism in the Darjeeling Hills: An Inquiry." *European Bulletin of Himalayan Research* 58: 1–22.

8 Clim(b)ing Slow-Moving Structures in the Garhwal Himalayas

Ainslie Murray

The Way to the Mountains Starts Here[1]

Some years ago, before the dams, the floods, and the ongoing construction of new road infrastructure, I walked the traditional Hindu pilgrimage route called the Char Dham Yatra, high in the Garhwal Himalayas. The pilgrimage links the four high-altitude sources of the four sacred rivers of Hinduism in India: the Yamuna at Yamunotri (3185m), the Bhagirathi at Gangotri (3415m), the Mandakini at Kedarnath (3584m), and the Alaknanda at Badrinath (3096m). In this chapter, I present this pilgrimage as an exercise in spiritual clim(b)ing. Paerregaard (2023) coined the terms "deep climing" and "social climing," with the latter pertaining to the socially mediated interactions of communities with their Deep Time environs, and changes therein. "Spiritual clim(b)ing," in this chapter, refers to both physical movement—namely, the climbing of geo-ecological earth architecture through pilgrimage—and the pilgrims' affective and spiritual interaction with the earth via the life-enabling sources of sacralized rivers and mountains. When framing pilgrimage as the affective, spiritually, and cosmologically entwined interaction of humans with earth architecture, agency, and life-sustaining affordances, pilgrims can be better understood as clim(b)ers. In this chapter, I interchangeably use the words "pilgrim" and "clim(b)er," but give more precedence to the latter when addressing the geo-ecological experience and affective perception of pilgrimage. Spiritual clim(b)ing is part of long anthropogenic history in the Himalayas, with sages, across religions, seeking quiet solace and spiritual realizations in the mountains. However, whereas spiritual clim(b)ing was earlier an exercise reserved for the few, today it attracts large numbers of spiritual practitioners and tourists. This, as I shall show, results in a paradox, as the geo-ecological structures and earth affordances that are being clim(b)ed for their sacredness are being anthropogenically affected as a result.

The physical climbing of the pilgrims is an ordered sequence of ascents and descents to four temples that mark the sources of the four rivers. From a clim(b)ing perspective, these four temples significantly reveal the human affirmation of the life-sustaining provisions of the four rivers, with pilgrimage becoming a celebration of their fecundity. The temples associated with the river sources are located downstream from the glacial sources, and are

DOI: 10.4324/9781003484394-8

accessible for six months each year before they are each enveloped by deep snow through the winter. Traditionally, pilgrims walked continuously for two months to visit the temples on a high-altitude route across the tops of the mountains. Today, the temples are linked by a network of precarious roads and paths clinging to unstable slopes. Large numbers of pilgrims arrive at the roadheads in vehicles and then disembark to continue their journey on foot. It is part austere journey and part measured mayhem as very large numbers of people set off into the mountains. Yangzom and Wouters (Chapter 10, this volume) invoke the notion of "mountain clime" to conceptualize how highlanders in Bhutan are simultaneously in the mountain and of it, acting in relation to it precisely as they are acted upon by it. In such a mountain clime, a mountain is therefore both "something" and "somebody." In this chapter, mountains do not primarily appear as geo-ecological structures people "live with," but rather as earth architecture humans move with, through, and in. Clim(b)ing as pilgrimage, indeed, invokes motion in mountains, as well as mountains in motion, because of both earth agency and anthropogenic impacts. The paths of the Char Dham route are remote and difficult to access. There are risks to health due to altitude. There are preparations to be made and there are things to be carried that enable pilgrims to climb and clime the architecture of the earth and the agency of the Garhwal Himalayas. The capacity of the pilgrims to survive in the mountainous landscape is complicated by the fact that they may be unfamiliar with the specific demands of the high-altitude clime. They arrive in a storied landscape to immediately emplace themselves in an environment that is at once illusorily familiar and entirely unknown. It is common to observe them walking into the mountains clad in thin cardigans and open shoes as they actively confront place and grapple with its climate, patterns, and agency through clim(b)ing.

Many pilgrims purchase guidebooks that contain current information about all aspects of the pilgrimage. With instructions on which prayers should be recited at which point, which items of clothing and equipment to bring, and current rates for mountain transport, these locally published books generally contain a list of recommended items for the pilgrim to carry (Pinkney 2014, 251). This list can be understood as an anthropogenic clim(b)ing guide and a microcosm of life's essential shelter and sustenance items—a collection of absolutely necessary provisions required to live for a short time in the mountains that is itself carried through the mountains. Along the pilgrimage route, these items are utilized in varying configurations according to the precise conditions of the body, the earth, the atmosphere, and the community. As these configurations are designed by each individual at each waypoint along the route, it's clear that the route is not only a line of slow-moving pilgrims on the path but also a line of slow-moving terrestrial landscape that reflects the broad interaction of social, ecological, and geological rhythms (Irvine 2020, 1). This line weaves together with the other patterns of the journey in quiet rhymes of footfall, place names, and modes of transport:

one / two / three / four
one / two / three / four / five / six
one / two / three / four / five / six/ seven / eight

del hi / harid war / janki chatti / yamu notri
del hi / harid war / janki chatti / yamu notri / janki chatti gan gotri
del hi / harid war / janki chatti / yamu notri / janki chatti gan gotri / gam ukh / gan gotri

air / road / road / foot
air / road / road / foot / road / road
air / road / road / foot / road / road / foot / foot

Walking slowly along this line (and it *is* slow for pilgrims unaccustomed to the altitude) with these items on the pilgrimage route enables an earth and anthropogenic architectural perspective on clime thinking. This perspective is developed from the mobility of the pilgrim across earth structures and the architectural structures they carry and make during the pilgrimage—structures of varying complexity that shelter and sustain the pilgrim on their journey—and how those structures interact with the evolving natural architecture of the Himalayas themselves. This thinking requires us to conceive of architecture not as a mute construct that is laid upon the equally mute earth, but as a changeable relationship of reciprocal influence as one creative force meets another in shared trajectories of profound durational contrast.

As Irvine (2020, 1) states, "There is nothing static about the terrain upon which we live and on which we depend." On the Char Dham route, the will of the pilgrim is in constant dialogue with the will of the Himalayas. This is not an easy relationship, and it is often characterized by dissonance as pilgrim and mountain come to terms with each other in volatile processes of growth and change. The infrastructure required to support the presence of large numbers of pilgrims is disruptive, often catastrophic, and always paradoxical, considering that reverence for river sources is what draws large numbers of people to the mountains. Conversely, the will of the mountains complicates the presence of the pilgrims despite their central role in the pilgrimage. Glacial retreat, declining water flows, and topographic instability due to development are daunting realities revealed by climate science that sometimes conflict with religious beliefs and the decisions that individuals make on a daily basis. Clime thinking on the Char Dham pilgrimage route offers a way of making sense of these conflicts by bringing together the "formerly parted domains between climate science and the lived earth" (Wouters 2023, 254).

The Himalayas are growing at a rate of five centimeters per year (Davis 2021, 224); changes in the landscape at that scale overwhelm our small steps within the mountains. Walking with others, with the things we need for shelter and sustenance that effectively emplace imagination, in a landscape that is in constant and sometimes dramatic and sometimes barely perceivable flux, is a direct experience that actively shapes our knowledge of the Himalayas through motion. It enables us to "pay attention to what lies beneath the ground" (Irvine 2020, 45) and through that attention to render geology locally meaningful as both motion and mountain guide how we live, temporarily, on the pilgrimage route.

Clim(b)ing is a productive way of thinking through earth and human architecture and agency that provides a manner of understanding sites in flux that questions the logic of measuring, surveying, and drawing places from a series of fixed and static viewpoints. Such thinking enables us to imagine beyond the present condition without freezing possibility into form (Rendell 2006, 188) and acknowledges that climes are changeable terrestrial places that are as much a part of us as we are of them (Wouters 2023). As "desire lines," the pilgrimage route mediates the desire of pilgrims to approach the river sources and the desire of the mountains to evolve at a pace and scale that radically exceeds the pilgrim. Our cartographic imagination is engaged as networks of lines traverse the surface of the earth, unfurling up and down and over and across—always responsive to surface topography and always locating us within the expansive horizontal world of the present, but never reflecting the timescales or related trajectories of the earth beneath or the atmosphere above (see Figure 8.1). If we shift from a horizontal to a vertical axis and reorient ourselves to imagine vertical "section lines" instead, we are

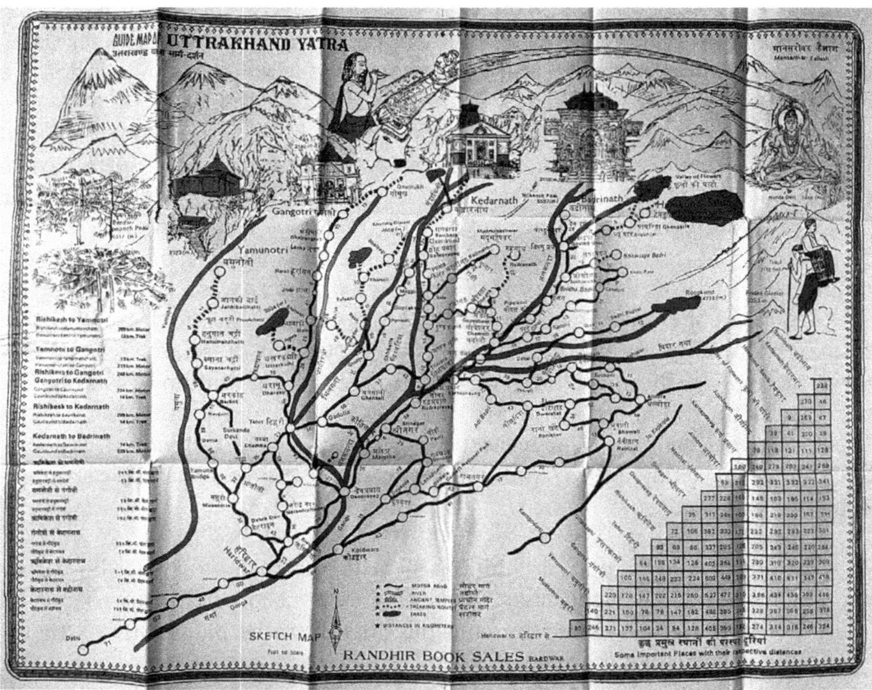

Figure 8.1 Char Dham Yatra guide map showing the 39 trekking routes
Source: Agarwala 1998 (Randhir Books, Haridwar)

suddenly compelled to think of multiple temporal and geological conditions simultaneously—a shift that greatly assists in further drawing together the domains of climate science and lived earthly experience. Walking alongside rivers in patterns of ascent and descent may assist us in developing a relationship of co-presence with the rivers and the environment (Ingold and Vergunst 2008, 7), and provides a glimpse into the Deep (geological) Time of the Himalayas. It relates a sense of deep climing, more-than-human climing, and spiritual clim(b)ing with the concurrent life-enabling and ever-moving forces of the earth. Our needs and our influence are dwarfed in this landscape that is far beyond us; yet, we must inhabit it and care for it with humility and diligence in our everyday acts of inhabitation and devotion. In essence, spiritual clim(b)ing reflects more-than-human curiosity and care for the life-sustaining provisions of Garhwal, even as this care is simultaneously offset by the anthropogenic impacts caused by the sheer numbers of pilgrims on the trails.

Pilgrim / Tourist / Adventurer

On the Char Dham route, people trudge upward on demanding ascents toward the river sources, driven by the goal of their destination and carried by other members of the community who share in the difficulty of the walk. The constant flow of people parallels the constant flow of the rivers below the paths. Increasing numbers of people on the route indicate the increasing diversity of visitors to the region; what was once the almost-exclusive domain of the pilgrim is now also attracting tourists and adventurers.

However, the vast majority of visitors to the Char Dham route are still pilgrims. To them, clim(b)ing is performed and experienced as *tirtha yatra* (literally, "undertaking a journey to river fords") and involves specific performative elements that impact the character of the journey (Das and Islam 2017, 243). The pilgrim undertakes the trek as part of a ritual and religious obligation and walks in a context of austerity, so that the earth's surfaces and the human body merge. Depending on how the pilgrim frames their pilgrimage, the journey can vary from a relatively straightforward walk in the mountains to ritualistic barefoot walking, walking on the knees, or rolling the body in the earth (Das and Islam 2017, 243). Most pilgrims walk ordinarily according to their own individual capacity and meet the challenge of this strenuous trek with a mixture of apprehension, exhilaration, and awe:

> Towering, snow-capped mountains hem in the rushing river whose roar as it crashes down the huge rocks of the riverbed often drowns out all other sounds. The air is thin, and pilgrims from the plains often struggle to catch their breath, exhilarated by the combination of majestic scenery and powerful religious devotion, not to mention by the experience of being in such a lofty and remote location.
>
> (Pennington 2015, 36)

The allure of a journey that promises a transition from contemporary inertia to energized consciousness (Rozelle 2015, 16) holds wide appeal, and pilgrim traffic has increased substantially in recent years. "Religious tourism" is experiencing rapid growth, and although it is not equivalent to pilgrimage in India (Das and Islam 2017, 244), there are fundamental similarities between tourists and pilgrims. Like the pilgrim, the tourist is also walking in a context that brings them beyond their usual habitat and into an environment of unknowns and uncertainties. We are reminded of Turner's observation that "a tourist is half a pilgrim if a pilgrim is half a tourist" (Turner and Turner 1978, 20), indicating that the line between pilgrim and tourist is indeed thin, perhaps growing thinner by the day, and that traditional dichotomies between pilgrimage and tourism may not even be appropriate in the world of post-modern travel (Das and Islam 2017, 242). Many people walking on the Char Dham route may be motivated half by pilgrimage and half by the immediate opportunities of education or recreation in a sublime environment. With increasing development to accommodate increasing numbers of visitors, improvements in the accessibility of the route prompt further collapse of the distinctions between tourism and pilgrimage.

The increasing numbers of clim(b)ers on the Char Dham route suggest that Himalayan pilgrimage in the modern era is a kind of religious tourism whose attractions include meaningful spiritual experiences woven together with a sense of adventure and of stepping outside the realm of daily life; it offers a chance to be among others in a geo-ecological world rich with stories and breathtaking scenery (Pennington 2015, 36) as well as a sense of the multi-species interrelatedness of gods, animals, plants, and people (Nath 2018, 425). Pinkney's survey of pilgrimage guide books suggests their changing content reveals declining differentiation between pilgrimage and tourism, or at the very least a change in how people conceptualize and access pilgrimage (Pinkney 2014, 255). The increasing availability of services to assist clim(b)ers, including various options for being carried to the shrines by another person or a group of people, pony, or helicopter, reflects the changes in the commodification of the trail and the general transformation of the sacred landscape into a sequence of religious tourism sites (Das and Islam 2017, 246).

Among the groups of pilgrims and tourists, there are also increasing numbers of adventurers who join the throng of clim(b)ers on the pilgrimage paths to access the landscapes irrespective of their religious significance. In the film *Meru*, which documents the first known ascent of Meru Peak, which rises just beyond the Gamukh Glacier, three elite mountaineers can be seen walking among pilgrims on the path from Gangotri (Chin and Chai Vasarhelyi 2015). They commence their journey as others do, amid the cacophony of pilgrims, ponies, and vehicles. Their attitude, as conveyed in the film, is one of singular focus and intent to "conquer" the peak, requiring a disposition quite opposite to that of the pilgrim or the religious tourist. The adventurer walks and climbs to surpass the environment, to "work" within it, to achieve, and to overcome; the journey is a specific and repetitious pursuit of awe and terror in the environment. For the pilgrim and the religious tourist, the experience of awe and terror, and the gradual dawning of the "ecosublime," may amplify their religious intent and lead to the integrated clim(b)ing that defines their experience.

This integration also operates reciprocally, where changes in the clim(b)ing population have prompted changes in the region—from changes in the amenity of the paths to broader changes such as the construction of the "all weather highway." Improvements to vehicular access promised through this substantial infrastructure project have caused significant damage, including considerable deforestation (*Times of India* 2023). While walking/clim(b)ing has been theorized as a way of achieving sustainable tourism due to its benefits to health, society, and environment, it is challenging to balance anthropogenic developments with the maintenance of authentic experience of the earth (Das and Islam 2017, 245). The catastrophic floods of 2013 and 2021 drew global attention to the inherent dangers of mass tourism in this remote region; with significant loss of life and many people still missing, the risks of the journey are exacerbated by climate change and rapid population growth (Pennington 2015, 43). Meanwhile, as Nath (2018, 420) observes, efforts to improve the infrastructure ultimately have an impact on the very core of the Char Dham Yatra:

> Improvements to the pilgrim road have come at a cost—the cost of pilgrims' phenomenological engagements with the landscape … [T]he experience of the sacred came from their ephemeral encounters with the natural world along the path to the shrines.

To convey a "thick" and "deep" sense of walking with architecture on the Char Dham Yatra, in which phenomenological experience is anchored in bodily action (Tilley 2010, 26), I will consider a series of architectural structures that emerge en route and interact with the geo-ecological earth architecture. These structures facilitate phenomenological experience and distill moments in which pilgrims touch and are touched by the Himalayas as they make and move between climes through the act of clim(b)ing (Smyer Yü 2015, 223).

Stick

The most prolific structures I encounter on the route are walking sticks. In the fluid throng of pilgrims, these sticks present additional complexity in the entanglements of ponies and palanquins. A stick is mentioned in every one of the lists of recommended *yatra* equipment I have encountered. They are variously referred to as "mountain sticks," "hiking sticks," or "walking/hiking/trekking poles." The majority of them are made of bamboo—rough hewn, practical, and plentiful.

Every stick functions as a structure to support the clim(b)ing pilgrim on the path. It is a vibrant extension of the body—a third leg to change weight distribution throughout the body, and to divert "work" from one part to another. A pilgrim leans on a stick with utmost faith: it pulls us up; it prevents us falling. We rely on the stick not only to mediate the physical strain of the walk but also to provide stability as we grapple with its psychological dimensions. The long and repetitive hours of difficult uphill and downhill walking are measured in the cadence of the stick; it becomes a kind of metronomic collaborator in both social and eco-geological domains. The stick supports the physical climb and simultaneously "sounds" the earth as a tool of correspondence between weather, terrain, society, and habitat—it calls attention to the terrestrial body and also to the rolling formation of climes as pilgrims move up and down the path. Following Ingold, the stick "impresses" on the earth as pilgrim and stick "go along together" (Ingold and Vergunst 2008, 8). Following Morris (2004, 131), the stick is a "climing" device that transfers knowledge between body and ground. With each impression, with each earthing, the stick enables us to "listen" to the mountains (Turpin 2013, 57). We listen to the dust, to the gravel, and to the layers of rock moving slowly beneath our feet. We lean on the stick and in response the earth asserts its deep time and pace; through the stick and with the stick as clim(b)ing device, we listen to the architecture of the Himalayas.

Chair

Many chairs and chair-like structures are used to carry clim(b)ers on the route. These range from beautifully made timber palanquins carried by four people with both rigid and suspended seats, to woven baskets worn on the back of a single carrier, to improvised slings carried by two people. These structures provide a valuable glimpse into mountain innovation in design and construction.

Some chairs appear to have been lifted quite literally from domestic settings simply by inserting two poles under the arms of a standard chair; others have been thoughtfully crafted with the comfort of both the carriers and the pilgrim in mind. These structures enable the pilgrim to face forward, looking ahead at the path to be walked. The number and organization of the carriers transform the walk into a procession that recalls the formal journeys of religious idols that are taken on palanquins to and from the temples each year. The environments of the pilgrim and the carriers are distinguished between the jointing methods of the structure. The pilgrim in a sling floats over the path as the feet of the carriers interpret and correspond with the natural architecture of the mountains; the pilgrim in a rigid chair feels every footstep of every carrier amplified through unyielding connections. Being carried through the mountains—a necessity for many pilgrims—may result in a more dissonant relationship with the earth as the rhythm of the earth is mediated and altered by the carriers.

Baskets, which may be woven locally, are generally used by very old or very young clim(b)ers. Rates for carrying them are based on body weight. The basket, which is padded with cloth at key points, is slung from the head across the carrier's back (see Figures 8.2 and 8.3). This carrying method positions the carrier and the pilgrim back-to-back, with the carrier facing the path ahead and the pilgrim facing in the opposite direction. In this manner, the pilgrim always looks backward, retrospectively considering the path. The structure flexes and creaks with each footstep, speaking of the mutual compressions of two moving bodies negotiating complex topography under significant physical strain. If the weaving of the basket captures a process of binding parts in sympathy through interstitial differentiation (Ingold 2015, 23),

Figure 8.2 Two pilgrims being carried in baskets on the path to Yamunotri in 2004
Source: Photograph by the author

Figure 8.3 Detail from Char Dham Yatra guide map showing a pilgrim being carried in a basket
Source: Agarwala 1998 (Randhir Books, Haridwar)

then the use of the basket in this manner creates a mutual shaping of the environmental experience in which the carrier, pilgrim, structure, and environment are intrinsically "knotted" together. This knot, as Ingold (2015, 23) explains, is "not a brittle one that allows for freedom only in the spaces left between, but a supple necessity that admits to movement as both its condition and its consequence."

Tent

The volume of pilgrims on the route places tremendous stress on the associated towns' infrastructure. Accommodation is in short supply in the pilgrimage season, and in remote Kedarnath and Bhojwasa (near the Gamukh Glacier), clusters of tents are set up to maximize the pilgrims' options. Shared tents provide clim(b)ers of limited means with opportunities to participate in the pilgrimage, so they play a vital role in ensuring relatively equitable access to the sites. The tents are also an important response to the seasonality of the pilgrimage and offer an economy of built structure that is commensurate with the patterns of inhabitation through the pilgrimage season.

Simple tents are set up in rows in close proximity to pilgrim rest houses, or independently by tourism organizations. They provide minimal shelter, offering a place of horizontal respite and basic protection after a day of walking. Clim(b)ers are provided with a cot and a blanket—a basic offering that accords with the austere character of the pilgrimage, in which attention is consistently drawn to what is "essential." The minimalistic accommodation plays an important part in the creation of a memorable experience, which can be attributed to the purposeful differentiation between *yatra* accommodation and a pilgrim's usual domestic setting.

Design theorist Vilém Flusser (1999, 57) describes how tents assemble experience, and how that experience is subdivided and diversified by the "screens" of the tent. Like the woven basket, the woven fabric of the tent's walls and ceiling creates a network of openness to wind and spirit that stores experience. In the tent, the pilgrim hears the wind ripping through the valley amid the exhalations of fellow pilgrims, whispered prayers, family chatter, and anticipation of the days ahead. It may even be possible to hear the slow grind of the moving earth.

Clothing functions in a similar manner to progressively adapt to and reflect local conditions. As the pilgrimage progresses through ever higher altitudes, pilgrims add layers of clothing to protect themselves. Each layer is an acknowledgment of and response to the immediate conditions. Therefore, clothing and the decisions made around its use may be understood as an assembly of experience that, like the tent and the clime thinking it affords, reveals the multitude of climes on the pilgrimage route.

Shack

The skeletal frameworks of various shacks perch precariously along the edges of the path. These are basic structures that cling to meager patches of horizontal earth—timber posts and beams lashed together with rope and covered in motley collections of sheet metal, textiles, and tarpaulins.

The beginning of the season sees a return to structures left dormant through the winter and the recommencement of enterprise and exchange. Clim(b)ers stagger in to perch on low stone walls, where they regain an even breath and the courage to persist through the receipt of sustenance. This pattern of receipt is shared beyond each individual to other pilgrims and other groups over days and years. The repetition of this moment across all walkers on the route on all days in all seasons is itself a space of shared understanding and possibility. The communitas that Turner (1969, 360) describes emerges in the repetition of the arriving pilgrim and the sharing of tea over and over again, placing the structure in a flow of perpetual coming and going along the rivers. The end of the season sees the occasion happen in reverse, as the shack is emptied of its pilgrims and its tea, and stripped of its cladding (see Figure 8.4). This pattern is effectively one of tying and then untying a knot at the beginning and end of the pilgrimage season, which, as Ingold (2015, 26) suggests, does not break things into pieces but is rather a "casting off" in which lines that were once bound together go their separate ways. The surface architecture comes and goes while the architecture of the Himalayas persists, independent of human perception and activity. As Rozelle (2015, 38) notes, to isolate life on the surface of the earth is to limit our relationship with the moving earth and to ignore the rich and full cross-section of life that clime thinking affords.

Live / Dwell / Inhabit

Morris (2004, 5) observes that, within the Cartesian tradition, our inquiry into space and all the forces that have a bearing on it begins with a space

Figure 8.4 Skeletal shacks clinging to the side of the path to Kedarnath on the final day of the 2004 pilgrimage season

Source: Photograph by the author

already structured in geometrical terms and then considers, secondarily, how the body operates within it. This model, put to us by mathematician–philosophers whose interests lie in the geometric reduction of space, relegates the body to a secondary or oppositional role (Morris 2004, 3). Not only are our bodies denied any meaningful role in the formation of architectural space, but the tradition actually conceals and suppresses the mutually affective encounter of body and earth.

In response, Morris (2004, 175) proposes a positive relationship in which the actions of our bodies are instead conceived as productive "crossings":

> Our bodies cross with the world, cross the earth, cross with our development and with our social world. Our sense of space refers to and makes sense of this crossing, it is not the reconstruction of an already constituted spatial order or container into which we have been dropped.

This crossing frames a relationship between body, earth, atmosphere, and community that is not simply the addition of one to the other but rather a sense of space that arises from the constructive interaction of all of these components. The body is not just loosely or non-specifically in the world; it is in a particular environment that is denoted by a precise confluence of conditions and trajectories. The body ascends the mountains and inhales ever-thinner air with and among other bodies; the air is exchanged through increasingly labored breath, progressively climing as the confluences of body, earth, atmosphere, and community form and reform with each footstep. Following Ingold (2011), these conditions and trajectories build a meshwork in which space is characterized by its myriad expansive and responsive textures rather than its fixed points.

As we move along the pilgrimage route, we generate strings of spaces that intersect with the world and other strings to form countless climes. Traces of these climes are found in "vibrant" materials (Bennett 2010) such as the visual string of footprints, in the wear of a stone seat, in the sheen of an often-touched handrail, and in a frayed piece of fabric. They are also present in changes in the air, displacements of dust, a dislodged leaf that was there and is now here, and in the evaporating moisture on the underside of an upturned pebble. A body that dwells habitually in space over time gathers these complicities and extends into a more complete picture of space. As each pilgrim makes and inhabits their own space, a collective temporal momentum emerges—a mobile architecture, a raft of architectures, a clime in motion within the moving mountains.

Are these spaces rivers? Are our bodies rivers? Are the moving mountains rivers? Until now, I have been walking beside the rivers, and crossing them en route to their sources, but am I already there, everywhere, through water, as the flush and flow of my own body connects me to others, to the earth, and to worlds beyond (Neimanis 2017, 2)? Rivers and water challenge us to understand situations and relations beyond the previously established logic of architectural space (Chen, MacLeod and Neimanis 2013, 277). What form do the crossings take if the structures I have discussed, which momentarily contact the slow-moving earth, become rafts, driftwood, or bubbles? How do we "measure up" the space of human inhabitation (Goetsch and Kakalis 2018, viii) at each moment that our structures are temporarily pinned to the earth? The Deep Time of the earth is not the background or mere incidental site of our endeavor—it is the architecture to which we respond with our own acts of emplacement and displacement. Our challenge, as Irvine (2020, 56) eloquently puts it, is "to ground ourselves, knowing that the ground is shifting."

I Walk on the Land to be Woven into Nature[2]

Feldhaus (2003, 18) points us to the geographical imagination and observes that rivers allow people to bring together spatially separated places in their imagination. This imagination, when unraveled across all rivers as a vast series of relationships, allows us to conceptualize as a whole the land across which the river flows. One part of a river is not seen as separate from another part because the river itself is moving. This movement is recognized on the Char Dham Yatra not only in consideration of the rivers themselves, but through the body as it moves alongside the moving rivers that themselves flow through the moving mountains. As the clim(b)er moves alongside the Yamuna, the Bhagirathi, the Mandakini, and the Alaknanda, steadily approaching their sources, the fluid material connection between many places prompts us to imagine others.

This imaginative and literal understanding of connectivity across places is vital in locating individual conceptualizations of how we inhabit the world and what it means to walk with and among slow-moving structures. It challenges any attempt we may make to inhabit the world in isolation and instead prompts us to see ourselves as part of a complex, more-than-human system that encapsulates physical and intangible matter as well as the union of physical and spiritual worlds. It follows Fulton's sensibility, of walking on the land to be "woven into nature" not through any overt documentation, physical creation, or achievement, but rather through accessing a more persistent kind of "inhabitation" through action that is itself a creative product of walking within cultural, ecological, and geological worlds. Through climing, we are able to see ourselves not as static points within a network, but as fluid, changeable lines that carry our intention and multidimensional aspects of inhabitation. This creative knowledge reflects a deep and instinctive knowledge of inhabitation akin to Ingold's (2011, 63) notion of "meshwork"—the wonderful, impossible entanglement of individual lines of being that productively extend and knot together with each other and the architecture of the earth. The architectural meshwork of our everyday lives that take place on the surface of the earth is intrinsically connected to the Deep Time of the earth through the stick, the chair, the ledge, the tent, and the shack, and we, as humans attending to the daily architectures required to walk the Char Dham route, are necessarily attending to and contributing to other life processes with different temporal spans that are hidden beneath our feet (Irvine 2020, 49).

Every clim(b)er on the Char Dham route is constantly making "architecture" by bringing structure to their walk and thus to their inhabitation of the route. Every clim(b)er makes daily decisions about which recommended items they engage with and in what combination, as well as how they engage with other architectural structures (stick, chair, tent, shack). In a sense, although the clim(b)er is walking with very little of the "architecture" they may normally expect to provide shelter and sustenance, they are also walking with everything they need. They have articulated for themselves a minimal threshold of "architecture" and it is not necessarily fixed, stable, material, or made by others. It is as mobile as they are, and it responds to the specific conditions of the moment. This construction, this clime, is actively assembled and disassembled every time the pilgrim pauses and sets off again.

Slope / Path / River

I walk on small stones that shear against each other underfoot, releasing clouds of gray Himalayan dust that feather and trail in my wake. I move within a rhythm, and project myself into that on which I tread. I am small, smooth, white. In the melee of passage, I am expelled into thin mountain space, tumbling at tremendous speed in the frictionless air until I am swallowed whole in one silent and invisible moment by the green, gurgling Ganges. I am engulfed, enveloped, transported. I am carried at a different speed, a capillary speed, a seeping, flowing, inevitable movement governed by topography and gravity. I am lush foothill, parched plain, saturated delta, muddy mouth. This walk connects me to all the places where the river flows (Feldhaus 2003, 42).

I clim(b)e and leave footprints—tidy trails that tell me where I am, where I have been, and possibly where I am going. Each compression of earth is a site of transmission between body, earth, atmosphere, and community. The compression is perceived in luminosity, in golden beams of light that emanate from my footprints into space (Jain-Neubauer 2000, 76). Time slows and extends as I consider my own tracks; it contracts, intensifies, collapses. Both footprints and light are within a single continuum, simultaneously visible and inhabited, luminous and illuminating. I am everywhere, altogether, a weaving, woven being whose passage is marked by disappearing impressions in the dusty gravel.

On the pathway to Gamukh, there is a particular moment that punctuates the trance-like walk of tired limbs and laboring lungs. On the path high above the river, we are brought to a halt by our guide, who directs our attention to the various signs describing the dangerous stretch ahead. Rather than being visible as a distinctive line within the landscape, the pathway disappears into a vast field of scree. We must not stop on this section. We must walk in single file, and quickly. Above, the mountain appears as a kind of edgeless fluid—a vast, vertical landscape of ominous potential. Below, the white waters of the river flow wildly. The imaginative grasp of the possible scale and consequence of a landslide is instantaneous, and changes the tone of our relationship with the ground.

Here, the footprint is no longer an affirming impression; it becomes instead a powerful condensation of uncertainty and risk. The more pressure that is exerted through the foot, the more perilous the situation becomes. This crossing is yet another texture in the meshwork and one that reveals the multiplicity of dependencies between body, earth, and environment. This moment can be understood as a kind of section through contrasting time-scales. Like the architectural section, in which the relationship between the inside and the outside is brought together in an impossible viewpoint, the act of walking on the path in this way makes explicit the intersection of scale and proportion, sight and view, and touch and reach that is rendered visible only in the vertical dimension (Lewis, Tsurumaki, and Lewis 2016). Clim(b)ing the path, in a state of emotional and possibly spiritual evolution, and simultaneously looking ahead to the unstable path, upward to the unstable mountain, and downward to the unstable river, I cannot help but think that this moment of embodied experience is tangibly entangled with the vast and unknowable movement of the earth. This is a moment when geological time intersects with the body and close-up embodied experience, a moment in which the specificities of earth, structure, inhabitation, atmosphere, and lived experience knot together as a primal comprehension of forces that usually elude comprehension due to their scale. In Irvine's (2020, 131) words, "Geological time here is something encountered within the time of autobiography, the vast story of environmental change intersecting with particular times and places in the course of life."

Dodging rockfalls from above while rapidly moving forward as the path behind slips down the mountain and into the moving river, the pilgrim negotiates challenging crossings with every step. The places of *tirtha* on the Char Dham Yatra are the "crossing places" and "fords" in which pilgrims may safely cross over to the far bank of a river, or to the far shore of the worlds of heaven (Eck 1981, 323). The *tirtha* is the point of departure, but conversely also the point of landing (Eck 1981, 326). It refers not to the goal, but to the way; not to the structure, but to the passage and the path one travels (Eck 1981, 325). The switch between approaching and arriving, or crossing and landing, reflects the validity of thinking of the pilgrimage route through knots, where things meet not "face-to-face, on the outside, but in the very interiority of the knot" (Ingold 2015, 22). Here again we can return to Morris and see our crossings as what the world affords in our movement and our inhabitation: "The body, I shall say, earths itself, using earth as a verb, on analogy with the verb ground, as when we say that something grounds a circuit" (Morris 2004, 132).

These observations of informal structures of the Char Dham Yatra have enabled me to explore some aspects of climes and clim(b)ing through architectural space and its production that impact on the way we see ourselves inhabiting the greater architecture we inhabit—the affective and relational architecture of the earth. Our daily lives are structured according to decisions we make around shelter, sustenance, and care, and those structures are of our own making. We walk alongside rivers—*with* rivers—on unstable terrain in the moving mountains in a complex choreography of interwoven, intentioned trajectories at vastly different scales. With every footstep, with every contact with the ground through a walking stick, a chair, a tent, or a shack, we "earth" ourselves and forge a relation to earth such that it is counterpart to our inhabitation (Morris 2004, 133). Different conditions demand different approaches to "earthing," and some demand a particular courage in how we relate our own journeys to those of the earth itself. In the face of changing climes and how we are to live within and through them, how we orient our own acts of shelter and sustenance on the Char Dham Yatra within that textured earthly meshwork is an individual and a collective responsibility.

Walking Walking, too Much Thinking Thinking[3]

It is the first part that frightens me, the first part of every walk in these mountains. I stand at the bottom and look up. For all the romance of walking the pilgrimage route, there is still an outrageously steep series of switchbacks that must be negotiated. Battled. And it *is* a battle, a physical battle and a psychological battle—body against mountain, body against mind. This first set of switchbacks—just the first set—requires me to climb higher than the highest mountain in my homeland. I can barely process it, but I make a start, knowing that it will be hard, and that I will want to stop again and again.

On the first few switchbacks, I feel fine— even strong—and I try to keep hold of the big picture. But the courage required to make the ascent soon dissipates as I face another turn, then another, and another. Mountains are confounding for flatland people. I cannot comprehend how big these mountains are and why I would want to clim(b)e them. They're big mountains. They're not particularly big mountains. The snow is deep. The snow is not that deep. It's cold. It's not that cold. I am alone. I am not alone. I have nothing I need. I have everything I need. I affect nothing. I affect everything.

Notes

1 Tufnell and Wilson (2002, 7) quoting Hamish Fulton.
2 Tufnell and Wilson (2002, 7) quoting Hamish Fulton.
3 Tufnell and Wilson (2002, 7) quoting Hamish Fulton.

References

Agarwala, A.P., ed. 1998. *Garhwal "The Dev Bhoomi": Step by Step Details of Char Dham Yatra – Thirty Nine (39) Trekking Routes with Multi-Coloured Map.* New Delhi: Nest and Wings.

Bennett, Jane. 2010. *Vibrant Matter: A Political Ecology of Things.* Durham, NC: Duke University Press.

Chen, Cecilia, Janine MacLeod, and Astrida Neimanis, eds. 2013. *Thinking with Water.* Montreal: McGill-Queen's University Press.

Chin, Jimmy and Elizabeth Chai Vasarhelyi, directors. 2015. *Meru* (film). New York: Little Monster Films.

Das, Subhajit and Manirul Islam. 2017. “Hindu Pilgrimage in India and Walkability.” In *The Routledge International Handbook of Walking*, edited by Michael Hall, Yaelm Ram, and Noam Shoval, 242–250. Abingdon and New York: Routledge.

Davis, Alexander. 2021. “Transboundary Environments, Militarisation and Minoritisation: Reimagining International Relations in the Himalaya from Ladakh, India.” In *Environmental Humanities in the New Himalayas: Symbiotic Indigeneity, Commoning, Sustainability*, edited by Dan Smyer Yü and Erik de Maaker, 220–238. Abingdon and New York: Routledge.

Eck, Diana. 1981. “India's ‘Tirthas’: ‘Crossings’ in Sacred Geography.” *History of Religions* 20(4): 323–344.

Feldhaus, Anne. 2003. *Connected Places: Region, Pilgrimage and Geographical Imagination in India*. New York: Palgrave Macmillan.

Flusser, Vilém. 1999. *Shape of Things: A Philosophy of Design*, translated by Anthony Mathews. London: Reaktion Books.

Goetsch, Emily and Christos Kakalis, eds. 2018. *Mountains, Mobilities and Movement*. London: Palgrave Macmillan.

Ingold, Tim. 2011. *Being Alive: Essays on Movement, Knowledge and Description*. Abingdon and New York: Routledge.

Ingold, Tim. 2015. *The Life of Lines*. Abingdon and New York: Routledge.

Ingold, Tim and Jo Vergunst, eds. 2008. *Ways of Walking: Ethnography and Practice on Foot*. Aldershot: Ashgate.

Irvine, Richard D.G. 2020. *An Anthropology of Deep Time: Geological Temporality and Social Life*. Cambridge: Cambridge University Press.

Jain-Neubauer, Jutta. 2000. *Feet and Footwear in Indian Culture*. Toronto: Bata Shoe Museum Foundation.

Lewis, Paul, Marc Tsurumaki, and David J. Lewis. 2016. *Manual of Section*. New York: Princeton Architectural Press.

Morris, David. 2004. *The Sense of Space*. New York: State University of New York Press.

Nath, Nivedita. 2018. “From Pilgrim Landscape to ‘Pilgrim Road’: Tracing the Transformation of the Char Dham Yatra in Colonial Garhwal.” *Journal for the Study of Religion, Nature and Culture* 12(4): 419–437.

Neimanis, Astrida. 2017. *Bodies of Water: Posthuman Feminist Phenomenology*. London: Bloomsbury Academic.

Paerregaard, Karsten. 2023. “Climing the Andes: Vertical Complementarity, Transhuman Reciprocity, and Climate Change in the Peruvian Highlands.” In *Storying Multipolar Climes of the Himalaya, Andes, and Arctic: Anthropogenic Climate and Shapeshifting Watery Worlds*, edited by Dan Smyer Yü and Jelle J.P. Wouters, 53–68. Abingdon and New York: Routledge.

Pennington, Brian. 2015. “Hinduism in North India.” in *Hinduism in the Modern World*, edited by Brian Hatcher, 31–47. Abingdon and New York: Routledge.

Pinkney, Andrea. 2014. “An Ever-Present History in the Land of the Gods: Modern Māhātmya Writing on Uttarakhand.” *International Journal of Hindu Studies* 17(3): 229–260.

Rendell, Jane. 2006. *Art and Architecture: A Place Between*. London and New York: IB Tauris.

Rozelle, Lee. 2015. *Ecosublime: Environmental Awe and Terror from New World to Oddworld*. Tuscaloosa: University of Alabama Press.

Smyer Yü, Dan. 2015. *Mindscaping the Landscape of Tibet: Place, Memorability, Ecoaesthetics.* Berlin and Boston: De Gruyter.

Tilley, Christopher. 2010. *Geologies, Topographies, Identities: Explorations in Landscape Phenomenology,* Volume 3: *Interpreting Landscapes.* London and New York: Routledge.

Times of India. 2023. "US President Biden and PM Modi's Joint Press Conference: Full Text." June 23. https://timesofindia.indiatimes.com/world/us/us-president-biden-and-pm-modis-joint-press-conference-full-text/articleshow/101206538.cms.

Tufnell, Ben and Andrew Wilson. 2002. *Hamish Fulton Walking Journey.* London: Tate Publishing.

Turner, Victor. 1969. *The Ritual Process: Structure and Anti-Structure.* Chicago: Aldine Publishing.

Turner, Victor and Edith Turner. 1978. *Image and Pilgrimage in Christian Culture.* New York: Columbia University Press.

Turpin, Etienne, ed., 2013. *Architecture in the Anthropocene: Encounters among Design, Deep Time, Science and Philosophy.* Ann Arbor: Open Humanities Press.

Wouters, Jelle J.P. 2023. "Conclusion: Multilateral Clime Studies." In *Storying Multipolar Climes of the Himalaya, Andes and Arctic: Anthropocenic Climate and Shapeshifting Watery Lifeworlds*, edited by Dan Smyer Yü and Jelle J.P. Wouters, 253–272. Abingdon and New York: Routledge.

9 The Geopolitics of Riverine Climes in the Eastern Himalayas

The Brahmaputra–Yarlung Tsangpo and the India–China Border Conflict

Alexander E. Davis

Introduction: Climes and Geopolitics in the Anthropocene

The Brahmaputra River today crisscrosses numerous anthropogenic borders as it flows from Tibet, down through the jungles and floodplains of Arunachal Pradesh and Assam, and drains into the Bay of Bengal through Bangladesh. Even this formulaic description, without which it is exceedingly difficult to give the river's location, carries with it the assumption that the river crosses the anthropogenic borders, not the other way around. This is reflected in dominant scholarship that near exclusively privileges state views of the river rather than, as this chapter attempts, river views of the state. Due to its ice, altitude, dense jungles, sediment, and floods, the Brahmaputra has been difficult for external political entities, be they empires or postcolonial states, to redirect and appropriate for human-only purposes. However, in the "New Himalayas"—the period in which human activity, including the creation of borders, is fragmenting the eco-climatic connections of the mountains—the Brahmaputra has become a target for geopolitical actors which seek to control water flows, territory, and resources (Smyer Yü 2021).

Throughout this book, we examine "climes" as the mutually embodied relations of climate, nature, culture, and place, looking at how people "clime" and "re-clime" geo-ecological gradients and niches in the Anthropocene context. Following Karsten Paerregaard (2023), we examine how people clime mountains through affirming their affective qualities and engaging in symbiotic exchanges with the deities that inhabit them (see Chapter 10, this volume). We also explore how people re-clime in an era of climate change, as ecosystems, peoples, and weather patterns shift (see Chapters 3 and 7, this volume). Clime is more helpful to us than climate in the context of the Brahmaputra because it draws our attention to the ways in which human activities are entangled with the region's ecologies, and the place- and (in this chapter specifically) river-making relationships between weather, climate, and place (Carey and Garone 2014, 284). It also draws our attention to the more-than-human communities and connectivities of the region, including its hydrological and ecological systems and its indigenous animal worlds (whose presence in and around the Brahmaputra generally preceded those of

DOI: 10.4324/9781003484394-9

humans). However, the re-climing—both driven by anthropogenic interests and impacted by the Anthropocene—of this region is being done not just by local, multispecies communities of life adapting to changing waters and climates, but also by geopolitical actors, particularly India and China, and their associated militaries.

This chapter integrates the field of international relations with Himalayan clime studies and environmental humanities, and brings questions of anthropogenic politics into conversation with the Anthropocenic Himalayas and its life-enabling environmental flows of water. It recognizes not only that political attention has scarcely been attuned to the great environmental challenges unfolding in and beneath much of the Himalayas, especially in relation to its rivers, but also, in turn, the impact of anthropocentric waters and climates on present and future Himalayan geopolitics. Specifically, I ask: what does becoming a geopolitical "hotspot" do to a clime? Similarly, how does clime hold a form of agency in a region's geopolitics? The transboundary Brahmaputra River Basin is an ideal place to think through these issues, as we see the gradual and interlinked transformation of its climate, nature, and culture amidst geopolitical disputes. Here, the river's indigenous flows, as created by the Deep Time agency of the earth, are being disrupted for the purposes of national development projects, military infrastructure building, and national security, as both India and China try to climate-proof and waterproof their economic growth and national security.

In this chapter, I examine an unfolding clash in which the Deep Time riverine clime of the Brahmaputra is being transformed by an emerging anthropogenic geopolitical clime. As a result, what we find in this river system is a multispecies geopolitics, in the sense that anthropogenic geopolitics is altering and reorganizing the dense mesh of networks of geological and ecological relations. The nature of this emerging geopolitical clime is partially captured by "Anthropocene geopolitics," discussed below. These are not separate climes. My conception of the riverine clime of the Brahmaputra is defined by the deep historical connectivity between the ecologies, cultures, and creatures (at once atmospheric, terrestrial, and aquatic) of the river system (Iqbal 2021, 105). The emergent geopolitical clime, however, is a transformation to the riverine clime that began developing when empires sought to border the river and mediate and transform its indigenous flows. It settled when postcolonial states began solidifying these claims. The Brahmaputra is being engineered, its ecosystems sliced up into bordered, state-based entities, split between India and China, and, to a lesser extent, Bangladesh and Bhutan. Here, we see state actors transforming the river system based on their mutual enmity while damming and transforming the river to fit geopolitical and economic state projects. This brings with it the often-imagined threat of a "water war" between India and China (Chellaney 2013, 309).

My question then becomes: can geopolitics re-clime the river, or will the riverine clime successfully re-clime geopolitics?

Drawing on a more-than-human geopolitical theorization, I survey the border history of the region and its relationship with the river. I then examine contemporary geopolitics between India and China. Here, I argue that the riverine clime continually shows its power by resisting the geopolitics, which affirms the river's refusal to become "modern" and its continuing agential powers. By repositioning the geopolitics of the region as being not just between India and China but also between humans and the earth, I expand the politics of the region by situating it in a multispecies frame. This shows us the more-than-human geopolitics of the Brahmaputra. Although I survey the broader river basin, my focus falls primarily on the stretch of the Brahmaputra near the great bend, looking at India's and China's approaches to dams, their contested border, and the role of the transboundary Adi community, including their companion species.

I conclude that relations between humans and the earth in this region are so close that the river will ultimately wash away the geopolitics, or the geopolitics will destroy the river.

More-than-Human Geopolitics in the Anthropocene

When we look at the issues faced by the Brahmaputra, interlinked threats of dams, climate change, multispecies and cultural loss, we can see entanglements between its riverine clime and the geopolitical situation. Classical geopolitical thinking examines the environmental context as a way of thinking through strategy, and noting which societies excelled and which failed, in an environmentally determinist fashion. However, today, as Simon Dalby (2020, 8) puts it when coining the idea of Anthropocene geopolitics:

> It no longer makes sense to see the world just as an external backdrop to the human drama, or a source of resources and a sink for wastes. The Anthropocene brings an end to these distinctions of nature and humanity. We live in an increasingly artificial world in which the choices are between a reasserted politics of dominance with increasingly militarized borders, or comprehensive attempts at economic innovation which recognize that policies of separation, and the invocation of sovereignty as a rationale for evading responsibilities across borders, are untenable.

Although much of this applies to the Brahmaputra River, the sense of this world being artificial is misplaced. It is more useful to think of geopolitics in this region as more-than-human, because the region, in its varied ecologies and multispecies networks, persistently resists "taming" by humans. The mountains and the river themselves played key roles in creating these more-than-human borders (Ozguc and Burridge 2023; Davis 2023). As we will see, attempts to engineer the region have resulted in numerous environmental problems, fractured governance, and local resistance, which have historically slowed attempts at infrastructure development. Historically, floods and earthquakes slowed colonization, made it difficult for external actors like the Mughals and later European colonial powers to survey, and still today resist attempts to dominate the river (Cederlöf 2013, 22–24). The recent devastating floods in Assam, despite decades of attempts to "control," "manage," or "govern" the river system, demonstrate its continued resistance. This is the riverine clime's geopolitical agency, its ability to resist human domination. To take this a step further, the riverine clime's agency is an inherent part of the earth's agency, which is still strongly felt in the region.

This is a form of more-than-human geopolitics, in which geopolitics is not just between competitive state actors but simultaneously between humans and the earth. In a fractious political borderland like the Brahmaputra River Basin, which is of enormous ecological importance because it nourishes multispecies communities, humans included, as well as geopolitical importance, this points us theoretically to the need to understand the relationship between geopolitics and ecology, humans and the earth.

International relations theorizing has generally looked at the Himalayas as a site of geopolitical contest in which the actors are the great powers of the region—India and China (Davis et al. 2021). When seen from a more-than-human geopolitics perspective, however, the contest for the Brahmaputra–Yarlung Tsangpo starts to look different. Rather than a geopolitical struggle between India and China, it becomes an anthropogenic struggle over a border, where the chief combatants are the armies and engineers of India and China and the river system itself. In particular, this case draws our attention to the power of water. As Irene J. Klaver (2022, 124) puts it, water "challenges clear-cut divisions and oppositions, undermines categorizations, messes up lines of separation, laughs at institutions, builds and resists infrastructures. It leaks, overflows, erodes, spreads, disappears, dilutes, and pollutes." It is this agential power that makes India's and China's efforts unlikely to succeed.

Riverine Clime, Geopolitical Clime

The broader Brahmaputra River Basin is part of what Smyer Yü (2021, 9) has called the "terrestrial ocean of the Himalaya," in which "water fully saturates land and land entwines with water." The Brahmaputra's headwaters begin at 5000 meters above sea level, near the Angsi Glacier. The highest peak in the river basin, Mount Kanchenjunga, reaches 8586 meters. Just 115 kilometers from this peak are sections of the river basin on the lowland plains of India (Gamble 2019). The Brahmaputra is one of the most sediment-rich rivers in the world, making the plain particularly fertile (Thomas 2017). However, due to extreme rainfall, the river and its tributaries flood regularly with spectacular intensity (Saikia 2019, 1–16). With the climate changing, these floods are expected to become more common and more intense.

The watershed is also intimately connected to coastal areas through river deltas and the monsoon. The Himalayas knit together mountains, oceans, and alluvial plains in climate-making processes (Chapter 1, this volume). The river is fed by Himalayan snowmelt in the dry section of Tibet, and this is the main feeder of the river during the dry season (Mohammed et al. 2017, 169). The eastern part of the basin, however, receives exceptionally high rainfall during the monsoon season, contributing substantially to its waters within India and Bangladesh. The watershed, then, also relies on the monsoon system, in which hot, tropical winds come up through the Indian Ocean and hit the high peaks of the Himalayas, resulting in rain that falls over South and Southeast Asia (Amrith 2013, 265). Hence, the ocean and the mountains are connected through global climatic systems.

The upper reaches of the river basin have not experienced high levels of human habitation, partly due to their ecologies. On the plateau, Tibetans have relied upon the river for water and irrigation for centuries, using it for agriculture through irrigated river flats. Similarly, the area of Arunachal Pradesh has been influenced by external actors for centuries through trade and pilgrimage networks. Both the British and Qing empires made claims over the region, without being permanently present (Gamble 2019). In Bengal, the East India Company, though able to profit, struggled to administer and control the region because it was out of step with the ecological and climatic conditions. Its officers found life particularly difficult in hill areas where they lacked local knowledge (Cederlöf 2013, 219–224).

Downstream, the multispecies worlds of Assam and Bangladesh rely on the river for water, agriculture, and livelihoods. Bangladesh relies on the Himalayan watershed and the silt it deposits throughout the country (van Schendel 2020, 1–5; Chapter 11, this volume). The silt is generated partly through the grating of the Himalayan rocks, as the South Asian continental plate crashes into Asia. Some 80 percent of Bangladesh lies in floodplains (van Schendel 2020, 9). The coastline around the Bay of Bengal is very prone to erosion at rates that might exceed the depositing of silt, potentially presenting an existential threat to Bangladeshi coastal and riverine communities (Brammer 2014, 51–62). However, annual floods also replenish the land, depositing silt. The communities of the river basin are very much adapted to patterns of flooding (van Schendel 2020, 7).

This is the Brahmaputra's riverine clime, created through millennia of Deep Time history. It is profoundly shaped by the nearby ice pack, the altitude, and the monsoon. Monsoonal rains mean the foothills of the river basin are particularly muddy, making it difficult to build major infrastructure. This is deeply entangled with its multispecies ecologies of life and the built environment (see Chapter 6, this volume). The river's history and ecology, though, have also been crucial to its political life. Through much of Tibet, the river has now been thoroughly engineered by the Chinese state (Gamble 2019). Once it leaves the high-altitude desert of Tibet and enters the jungles of northeast India, tumbling down sharply, it changes character as it gains more water from the monsoon. The exceptional rainfall of northeast India during the monsoon contributes to the river basin's waters, and makes the jungled foothills similarly difficult to traverse without local ecological knowledge.

Historically, local communities in northeast India and Bangladesh were used to the river's unruly nature and learned to live alongside its flooding and high sedimentation, though of course this was not without its risks.

This leads to the emerging geopolitical clime of the Brahmaputra. However, in order to understand the contemporary geopolitics of the river, and the contest for its waters, we should first investigate how the Brahmaputra became an international river. This draws us into the history of state-making in the region and points to a larger range of actors who are important in governing the river (Thomas 2017).

Ruth Gamble (2019) has examined the efforts of Qing China and British Indian explorers to map, explore, and control the upper reaches of the Brahmaputra River Basin. However, neither the British nor the Chinese permanently inhabited this area, and neither succeeded in asserting their authority on the ground. Borders were drawn on the supposed basis of "natural boundaries," such as watersheds and mountain ranges (Goettlich 2019). The McMahon Line between India and China, the site of today's border conflicts, was determined at the Shimla Conference of 1913–1914. The British, whose insistence on territorialization and defined borders had been the impetus for the conference, came prepared with a surveyor's map that marked the "natural" borders between Tibetan-ruled "Outer Tibet," Chinese-ruled "Inner Tibet," and British Indian territories in the western and eastern Himalayas. All these lines were drawn by British cartographers along watersheds and mountain ridge-lines (Davis et al. 2021). Such line drawings could not consider the river's intricate flows, let alone the cultural and kinship relationships of the region, or the yaks, elephants, and tigers, whose habitats and traditional migration patterns have today been disrupted by military infrastructure, fortress conservation, and border patrols (Davis 2023, 113–150). The region that the British claimed, particularly Tawang, had strong religious ties to Lhasa. For the British, it was unacceptable to have Chinese-controlled territory so close to the plains of British India (Gamble 2019, 52).[1] These are the origins of the river system becoming both international and a borderland.

With decolonization, India, China, and Pakistan began making territorial claims on the basis of their imperial forebears. These claims related not just to people and culture, but also to natural resources and environmental topography. China based its claims on a sense of Tibetan ethnicity, Pakistan claimed all of South Asia's Muslim-majority territory, while India claimed to be home to all of South Asia's religions. Therefore, these new states' territorial claims were based on the culture, religion, and ethnicity of local communities. For sparsely populated ice caps and the diverse Himalayan foothills, this was not a helpful organizing principle. As a result, borders were also drawn in relation to physical topography, as per the English "science" of border-making, but also with reference to resources, water, and ports, as well as high peaks and ice caps (Thomas 2017), thus fragmenting eco-climatic continuities and multispecies communities.

The Brahmaputra River Basin was made more "international" with the creation of independent India and its partition. The creation of East Pakistan necessitated forming borders. Again, British boundary-makers used rivers as territorial markers. Bengal was partitioned into West Bengal (India) and East Pakistan in 1947 partly so that natural resources, particularly rivers, water, and port access, would be shared (Thomas 2021). The river itself, then, was part of the border-making process. However, water situated between India and Bangladesh flows through no fewer than 50 different rivers, in what van Schendel (2020, 4) calls a "crazy pattern of channels, marshes and lakes." Partition created a series of serious political issues, with competing forms of nationalism (including Bangladeshi, Indian, Assamese, and Bodo, among others) all connected to the region's territory. The India–Bangladesh border was further demarcated and simplified in 2016 through the removal of 198 enclaves, with full citizenship granted to their inhabitants, demonstrating what van Schendel (2002) describes as the slow, ongoing process of border-making.

Since decolonization, states have gradually increased their control over the region and its waters. Initially, the states within the Brahmaputra River Basin sought the loyalty of its peoples (Guyot-Réchard 2016). India and China fought a brief war over the McMahon Line in 1962, with the Chinese occupying Tawang in Arunachal Pradesh and moving through river valleys connected to the plains of Assam (Gamble 2019). They eventually retreated back over the border, but continue to claim almost all of Arunachal Pradesh and parts of Assam as their territory.

As for the river itself, due to its sedimentation and flooding, the Indian government has repeatedly tried but failed to find ways to control and regulate its flow. Arupjoyti Saikia (2019, 475) summarizes these efforts as follows: "Several years of working with international experts led to one simple realization: there hardly existed any expert knowledge about the river and without this nothing could be done to tame it." For instance, American engineers' knowledge of US river systems did not aid their attempts to master the Brahmaputra. Nevertheless, as Saikia (2019, 476) puts it: "None agreed that their failed measures were based on a fundamentally erroneous understanding of the river's nature."

The use of engineers, cartographers, and exploration to determine borders, using the earth's water bodies and landforms, such as ice caps, mountains, ridges and rivers, shows the need to reconceptualize the environment's role in international politics. As Thomas (2017, 46) puts it regarding the use of the Ganga River as a border, "national borders and international rivers determine one another … [This] demands that we reconceptualize international rivers as synergistic, multifaceted and ongoing interactions between rivers and borders." Water flows are crucial to agriculture across the region. The Brahmaputra River's silt level makes it particularly fertile and difficult to dam. Similarly, its floods have meant that the region was seen by outsiders as difficult to control. In this sense, the river system is partially responsible for the location of the borders it traverses. Its more-than-human jungles, ice caps, and floods have contributed to its construction as a borderland.

Although the communities that live along the highest sections of the river basin have historically been spared substantial external control, today the situation has changed dramatically. Based partly on their dispute over Arunachal Pradesh, India and China have both sought to solidify their control and influence over borderland territories. These geopolitical rivalries etch themselves into the landscape through anthropogenic transformations. China has built infrastructure right up to the border, and even occasionally pushed across the Line of Actual Control (LAC) (Brethouwer et al. 2022, 3).[2]

Fitting these diverse multispecies worlds into the black box of a nation-state was never going to be easy. However, as we will see, given the dispute over the border between two vast postcolonial nation-states, the shift from colonial frontier to postcolonial borderland has led to an even more difficult political and ecological situation. It is here that the region's geopolitical entities, its nation-states, have begun to transform the clime of the Brahmaputra.

Damming the Brahmaputra River Basin: Geopolitics between Humans and the Earth

Dams are the physical manifestation of the region's re-climing, in the sense that they obstruct, contain, and manage indigenous environmental flows in an Anthropocene context. Gamble (2019, 2022) notes that the river has been thoroughly engineered in Tibet, with dams celebrated as providers of clean energy. In China, the Brahmaputra's sand is dredged and turned into concrete, which is then used to build the dams. Prior to this, China had already completed numerous dams on the river's tributaries, and there is a cluster of them at Zangmu, Dagu, Jiexu, and Jiacha. These dams attracted considerable media coverage and even panic in India, leading to the rumor that China was planning to withhold water or divert the river entirely (Deka 2021). Once it crosses into India, the Brahmaputra and its sediment-rich waters traverse jungle, sludge, and mud, all of which make building infrastructure much more complicated. Consequently, the Indian government has been much slower to dam the river system, despite estimates that the state of Arunachal Pradesh alone could meet one-third of India's electricity needs through hydropower (Mimi 2017, 219). Some dams have already been constructed in the state, and others are planned, but these projects have generated considerable local opposition.

Given the border conflicts, dams have become military assets. They represent massive, high-value infrastructure targets and so require substantial state security forces. Dam construction contributes to the securitization[3] of the Brahmaputra's waters, as it enables the diversion and storage of water resources, preventing them from flowing downstream. This is the logic of the emerging geopolitical clime—state and human control over waters based on borders, thus transforming indigenous flows for anthropogenic purposes. Dams' damaging local consequences often lead to local protests and resistance, which speak not only of their impact on humans but especially of the ways in which they denude ecologies and impede the flow of multispecies life. However, when infrastructure becomes a military necessity, it tends to be built far more quickly. Indeed, the rate of infrastructure development in northeast India has quickened in the past decade.

Military concerns and strategic logic have fed into the situation that the Brahmaputra faces today. Here, we see the region's riverine clime contesting with its emergent geopolitical clime. The state has explicitly recognized the Himalayas' agential geology as an obstacle to overcome when engineering the Brahmaputra on the Indian side of the border. The region's Deep Time history underpins its clime: the ongoing "rise" of the Himalayas as the Eurasian and Indian subcontinental plates collide leads to geological relationality—the rocks are grating against one another. This means that the mountains are in motion and prone to earthquakes (Gergan 2017, 490–498). The Himalayas are geologically young, then, which feeds both geopolitical and geological instability. One Indian engineer puts this down to the mountains being full of "geological surprises" (Gergan 2019). This curious turn of phrase does indeed see the mountains as having agential entanglements with anthropogenic projects, but positions their power as a problem for the state. The living mountains of the Himalayas need to be at worst subdued, at best tamed, by the state, but this may prove impossible. Landslides, mud, and heavy rain have long delayed the building of major infrastructure projects on the Brahmaputra (Saikia 2019, 491–492).

Both Indian and Chinese political elites have described the upper reaches of the Brahmaputra as a peripheral area—that is, one difficult to reach, govern, and dominate. They also view the people of the region as "backward" and in need of development, so justifying present and future planned anthropogenic transformations. China's model of governance has been far more efficient but far less responsive to local concerns.

In India, dams have faced quite substantial resistance from local communities, but are often constructed anyway (Gergan 2017). It is the national security approach to dam-building on the Yarlung Tsangpo–Brahmaputra between Tibet and Arunachal Pradesh that matters most for an investigation of its Anthropocene geopolitics. While both China and India have been building dams since the time of India's independence, it has taken the latter comparatively longer to build dams in the hills of the Brahmaputra, partly due to a lack of capacity in their borderlands (Gamble 2019).

The region's first large dam projects began operations in the early 2000s, and were immediately the subject of political wrangling over possible environmental consequences. In 2017, mud and concrete discharge from dams flowed down the river, sparking considerable local concern (Deka 2021). The dams also block silt flows, detracting from the river's ability to replenish agricultural land. Meanwhile, the accumulation of silt threatens regional dams' functionality. No trans-boundary environmental impact assessments (EIAs) have been performed over the Brahmaputra River Basin (Deka 2021); indeed, in times of conflict, China tends to stop sharing water-flow data with

India. The lack of trans-boundary EIAs is particularly concerning when we turn to India and China's current conflict over the region's waters, and especially the dam rush in Pemakö, where the river flows from Tibet to Arunachal Pradesh.

The Future Geopolitical Clime: China's Mega-Dam and India's Upper Siang Projects

Here, to draw out the connection between the clime, the river, and geopolitics, I explore India's Upper Siang hydroelectricity projects and China's mega-dam on the great bend of the Brahmaputra, examining their geopolitical meaning, their construction processes, local inhabitants' responses to them, and their potential environmental impact. These are aspirational projects. If successfully completed, they will dramatically transform the river system, including its hydrological systems and its multispecies climes. It is impossible to determine their full environmental impact until they are operational. Nevertheless, studying them now is worthwhile as doing so sheds light on the enormous ambitions of India and China, their ways of imagining the river system, and the patterns of thought behind the projects.

China's mega-dam will be located in Medog County in Tibet, the last county before the LAC. Unlike dams on the plateau, this one, at a lower altitude, will receive monsoonal rainfall, which could have a greater impact on the flow of the river than previous projects (Modak 2020). Though the plans are not yet finalized, the idea behind the dam is quite remarkable. There is a drop of over 2000 meters from one side of the great bend in the river to the other. This bend is situated between two of the tallest mountains of the eastern Himalayas—Namcha Barwa and Gyalha Peri. Here, the river flows through one of the world's deepest canyons. The planned Motuo Hydropower Station, rumors of which have been circulating since 2010 (Watts 2010), would rely on a water-diversion tunnel under Namcha Barwa (Doman, Shatoba, and Palmer 2021). This will require cutting a tunnel underneath the mountain to divert the water from one side of the great bend to the other, thus changing the biosphere from below. An access tunnel has reportedly been completed, suggesting that this approach to the dam is China's preferred method (*Xinhua News* 2021).

Medog County has a population of just 14,000 people, making it one of the most sparsely populated regions of the world. It is, however, home to a rich, biodiverse multispecies world that includes clouded leopards, pandas, and a tiny population of Bengal tigers—the only tigers in China (Yingqing and Liqiang 2019). The region was connected to the Chinese highway system only in 2013 (*Xinhua News* 2017), revealing how the process of colonization implemented by outsiders claiming to rule the region while living outside is still transforming the upper reaches of the Brahmaputra River Basin. Regardless of whether or not the local communities are displaced, the transformation of the Chinese side of the border, with a new emphasis on hydropower, tourism, and intensive farming, grants limited, if any, space for their connection to the land within the government's development plans for the region (Gamble 2022).

For example, Namcha Barwa is a sacred site for several local communities—the Adi, Pemaköpa, and Tibetans—and the mega-dam plan will require cutting into their goddess's heart (Gamble 2022). Damaging, summiting, or disrespecting these abode mountains has long been a cause for mass protest and anger directed at external actors (Wouters and Heneise 2022, 27–28). Local beliefs are based on centuries of habitation on both the land and the river. These subterranean earth forces are powerful, agential, and capable of completely reshaping the region's geopolitics. Yet, there is no space for the local community's cosmologies within the damming projects. Regardless, should the mountain decide to shake off the dams through natural disasters, as a manifestation of earth agency, it could well do so.

China claims that, as a "run of the river" dam, the flow of water downstream will not be impeded. Indeed, the head of China Power not only argues that the dam is a "project for national security, including water resources and domestic security" (Jie and Xiaoyi 2020) but insists that it will actually aid cooperation with downstream neighbors. The Chinese cite existing mechanisms, such as the Lancang–Mekong Cooperation organization, in support of this argument. However, the Lancang–Mekong River has been thoroughly engineered, dammed, and transformed into a human river (Geheb and Suhardiman 2019), so it is safe to assume that more of this form of state-based cooperation will lead to greater human dominance over nature. Hence, although it is often hailed as the key to avoiding "water wars," such cooperation might actually facilitate the demise of the riverine clime. Moreover, Lancang–Mekong Cooperation is dominated by China, so India views the prospect of a similar organization for the Brahmaputra as a threat to its control of the river.

The mega-dam project poses a particular risk to humans due to the threat of earthquakes (Huber 2019), which occur with considerable frequency, often reshuffling geo-ecological niches. However, our understanding of the region's tectonics are undermined not only by political tensions, which make research more difficult, but also by the earthquakes themselves, which have damaged or destroyed many archival sources (Bilham 2019). After reviewing the remaining historical and geological data, Bilham (2019, 474) concluded that, "distressingly, we are no closer to knowing where the next damaging earthquake will occur than two decades ago."

On the Indian side of the border, there are plans for a significant increase in the region's hydropower dams. As Sanjib Baruah (2017) notes, by 2011, the Arunachal Pradesh government had signed no fewer than 132 memoranda of understanding with developers for new hydropower projects. Few of these have been completed, but the figure nonetheless shows the speed at which the plans were drafted.

The Upper Siang project dates back to 2009, although its scope has changed several times since then. These changes are linked to ever more securitization, as China's dam-building projects have accelerated, and to local opposition, including from the area's Adi and Tibetans (Gamble 2019). That said, the project was securitized from the very beginning. Jairam Ramesh, one of India's more pro-climate-action environment ministers, decided to "fast track clearance of projects in the Siang river basin" due to "its strategic importance in the India–China border issue … [R]oad works along the India–China border should also be seen in this light" (Ramesh 2015, 170). In a letter to Prime Minister Manmohan Singh, Ramesh recounted a report from his secretary for the northeast that highlighted the security dimension:

> He said that in order to strengthen our negotiating position with China, overriding priority should be given to hydroelectric power projects on the Siang River. There are hydroelectric power projects on other rivers like Subansari and Dibang but from an international point of view, it is the Siang basin projects that are of strategic significance.
>
> (Ramesh 2015, 170)

Ramesh concluded:

> Clearly, we should take up projects on the Siang River basin as a matter of urgent priority. If this means even more attractive rehabilitation and resettlement (R&R) packages [for local people displaced by the new dams] and if this means giving additional incentives to the Arunachal Pradesh government, we should agree.
>
> (Ramesh 2015, 170)

The Siang project initially comprised two dams on the main channel of the Siang–Brahmaputra between Gelling and Pasighat. As part of this push to dam the river, a total of 44 memoranda of understanding were signed by the government of Arunachal Pradesh (Gamble 2019). Some of these related to small projects to power local communities. Others, like the Siang project, aimed to provide power to the whole of India. It is these larger projects that have attracted the most opposition, particularly from the Adi community, who accuse them of denuding human and other-than-human habitats. The largest Siang project has been delayed by a number of court cases, particularly as the dam's reservoir has been seen as an existential threat to the local community (Gamble 2022). Nevertheless, in 2017, the central government proposed replacing stages one and two with a single mega-project. In response, Tasik Pangkam, leader of the Siang Indigenous Farmers' Forum (SIFF), stated: "There is lurking danger of several of our small tribal communities getting wiped out from the face of the earth. Once uprooted, our culture, our language, our heritage will be all lost simply because some people elsewhere require electricity" (quoted in Kashyap 2017). He continued that the project threatened the very existence of the villages of Gelling, Tuting, Yingkiong, and Geku.

Yet, since then, the scale of the project has increased, with the plan now to build a 280–300-meter dam at Yingkiong that will store some 10 billion cubic meters of water (Baruah 2022). Director of the National Hydro Power Corporation Abhay Kumar Singh has argued that this dam will provide "green energy," control floods, develop fisheries, and act as a catchment for water. This has been driven by the fear that China might one day "open the gate" and flood India (Baruah 2022).

In 2019, there were some signs that the Arunachal Pradesh government might pull back from its large-scale hydropower plans, in favor of smaller, community-level projects (Karmakar 2019). However, ongoing concerns over China's plans has caused the national government to double down on the need for the Siang project (Gamble 2022). If both projects were to be completed, the level of hydropower extraction within this small, contested section of the river would transform indigenous water into anthropogenic water in a definite sense. India's anxiety about China diverting the river has led to the securitization of its waters and contributed to the dam-building rush. The sense of urgency articulated by Ramesh is now commonplace throughout India, while China's plan to dam the great bend is a dramatic escalation. These projects are inherently risky due to the likelihood of earthquakes, the presence of angry deities, the common unexpected consequences of dams, human error, and the displacement of people and other-than-humans with local environmental knowledge. Moreover, previous efforts to dam the Brahmaputra have largely failed to moderate floods and may even have exacerbated them (Akhtar 2017). Yet, the region's state actors still seem intent on building ever more dams.

International relations and state-based approaches do not consider the ecosystem and multispecies communities of life, which, along with humans, have been displaced by this region's transformation into an anthropogenic borderland. Moreover, they neglect the power of the hydrological system and the long-term consequences of disrupting indigenous earth flows. This must lead us to ask whether humans—or perhaps more accurately states—will ever actually control the river. Can this emergent geopolitical clime really control the Deep Time riverine clime of the Brahmaputra?

Conclusion: Re-climing the Brahmaputra?

The Brahmaputra River's experience suggests that there is a deeply entangled relationship between its nature, culture, and climate and geopolitics. Environmental factors were central to turning the Brahmaputra into an international river, a status the river itself resists by continuously interrupting and overpowering anthropogenic projects. Its ecology similarly led to the ongoing border conflict in the region. Its climate—especially the cold winters and muddy monsoon—has made it difficult for external actors to govern. The limited capacity of the British and Qing empires to govern the region was tied to their inability to maintain an administrative presence in the upper reaches of the river. This produced the contemporary border conflict between India and China. In the postcolonial period, the river basin's underlying ecology, with its silt, monsoons, floods, jungles, and community resistance, has limited humans' efforts to geo-engineer it. The riverine clime of the Brahmaputra has played a key role in the construction of its contemporary geopolitics, which have, themselves, begun to transform the river's clime.

The region's clime and geopolitics now find themselves in what appears to be an accelerating feedback loop. The bordering process has facilitated the transformation of the river, and the transformation of the river is exacerbating geopolitical tension over its waters. The region's clime has played a key constitutive role in it becoming a geopolitical hotspot, and the geopolitics are reshaping the clime.

This leaves us with the question of whether this particular river basin can be thoroughly reshaped by armies, engineers, and state elites. Other Himalayan rivers have been through this process—both the Yellow River and the Lancang–Mekong have been thoroughly dammed and geo-engineered for human use, for the purpose of development and modernity. The escalating dams-race between India and China, with limited cooperation over the river's main channel, is the key test case for whether humans will "win" the geopolitical struggle for the river. The engineering skills of the Chinese and Indian states are doubtless improving, as dams and their associated infrastructure move to ever-higher terrain. Over the longer term, the scale of the planned dam projects would genuinely transform the riverine clime of the Brahmaputra.

Continuing to transform the river through state-based extractive competition sets up a contest not just between India and China, but between humans and the earth. There is no "winning" in this adversarial contest. Ultimately, the river will wash away the geopolitics, or the geopolitics will destroy the river.

Notes

1 The bordering process on the plains of the Brahmaputra River system was similarly complicated, including the partition of Bengal in 1905, decolonization, and the creation of Bangladesh. Here, I focus primarily on the India–China border in the high Himalayas due to space constraints.
2 The LAC is the de facto international border between India and China, separating Indian-controlled territory from Chinese-controlled territory.
3 Securitization is the process through which an object or issue becomes considered as a national security threat to a particular state. For a foundational discussion, see: Balzacq (2011).

References

Akhtar, Mubina. 2017. "Dam Worsens Flood Devastation in Assam." *The Third Pole*, August 25. www.thethirdpole.net/en/livelihoods/dam-worsens-flood-devastation-in-assam/.

Amrith, Sunil S. 2013. *Crossing the Bay of Bengal: The Furies of Nature and the Fortunes of Migrants.* London: Harvard University Press.

Balzacq, Thierry. 2011. *Securitization Theory: How Security Problems Emerge and Dissolve.* Abingdon: Routledge.

Baruah, Rituraj. 2022. "Arunachal Dam Project May Cost India 1.13 Trillion Rupees." *Mint.* www.livemint.com/politics/news/arunachal-dam-project-may-cost-india-1-13-tn-11655746016904.html.

Baruah, Sanjib. 2017. "Whose River is it, Anyway? The Political Economy of Hydropower in the Eastern Himalayas." In *Water Conflicts in Northeast India*, edited by K.J. Joy et al., 116–144. Abingdon and New York: Routledge.

Bilham, Roger. 2019. "Himalayan Earthquakes: A Review of Historical Seismicity and Early 21st Century Slip Potential." In *Himalayan Tectonics: A Modern Synthesis*, edited by P. Treloar and M.P. Searle, 423–482. London: Geological Society.

Brammer, Hugh. 2014. "Bangladesh's Dynamic Coastal Regions and Sea-Level Rise." *Climate Risk Management* 1: 51–62.

Brethouwer, Jan-Tino, Robbert Fokkink, Kevin Greene, Roy Lindelauf, Caroline Tornquist, and V.S. Subrahmanian. 2022. "Rising Tension in the Himalayas: A Geospatial Analysis of Chinese Border Incursions into India." *PLoS One*, November 10. https://doi.org/10.1371/journal.pone.0274999.

Carey, Mark and Philip Garone. 2014. "Forum Introduction." *Environmental History* 19(2): 282–293.

Cederlöf, Gunnel. 2013. *Founding an Empire on India's North-Eastern Frontiers, 1790-1840: Climate, Commerce, Polity.* Oxford: Oxford University Press.

Chellaney, Brahma. 2013. *Water, Peace, and War: Confronting the Global Water Crisis.* Lanham: Rowman & Littlefield.

Dalby, Simon. 2020. *Anthropocene Geopolitics: Globalization, Security, Sustainability.* Ottawa: University of Ottawa Press.

Davis, Alexander E. 2023. *The Geopolitics of Melting Mountains: An International Political Ecology of the Himalaya.* Singapore: Palgrave.

Davis, Alexander E., Ruth Gamble, Gerald Roche, and Lauren Gawne. 2021. "International Relations and the Himalaya: Connecting Ecologies, Cultures and Geopolitics." *Australian Journal of International Affairs* 75(1): 15–35.

Deka, Bhaskar Jyoti. 2021. "Hydro-Politics between India and China, The 'Brahma-Hypothesis' and Securing the Brahmaputra." *Asian Affairs* 52(2): 327–343.

Doman, Mark, Katia Shatoba, and Alex Palmer. 2021. "A Mega Dam on the Great Bend of China." *ABC News*, May 25. www.abc.net.au/news/2021-05-25/chinas-plan-to-build-mega-dam-on-yarlung-tsangpo-brahmaputra/100146344.

Gamble, Ruth. 2019. "How Dams Climb Mountains: China and India's State-Making Hydropower Contest in the Eastern-Himalaya Watershed." *Thesis Eleven* 150(1): 42–67.

Gamble, Ruth. 2022. "Surviving Pemakö's Pluriverse: Kunga Tsomo, the Goddess, and the LAC." *Critical Asian Studies* 54(3): 398–421.

Geheb, Kim and Diana Suhardiman. 2019. "The Political Ecology of Hydropower in the Mekong River Basin." *Current Opinion in Environmental Sustainability* 37: 8–13.

Gergan, Mabel Denzin. 2017. "Living with Earthquakes and Angry Deities at the Himalayan Borderlands." *Annals of the American Association of Geographers* 107(2): 490–498.

Gergan, Mabel Denzin. 2019. "Geological Surprises: State Rationality and Himalayan Hydropower in India." *Roadsides* 1: 35–42.

Goettlich, Kerry. 2019. "The Rise of Linear Borders in World Politics." *European Journal of International Relations* 25(1): 203–228.

Guyot-Réchard, Bérénice. 2016. *Shadow States: India, China and the Himalayas, 1910–1962*. Cambridge: Cambridge University Press.

Huber, Amelie. 2019. "Hydropower in the Himalayan Hazardscape: Strategic Ignorance and the Production of Unequal Risk." *Water* 11(3): 414–436.

Iqbal, Iftekhar. 2021. "Rivers of Mobility: Multi-Ethnic Societies and Ecological Commons in a Fluvial Asia." In *Yunnan–Burma–Bengal Corridor Geographies: Protean Edging of Habitats and Empires*, edited by Dan Smyer Yü and Karin Dean, 105–121. Abingdon and New York: Routledge.

Jie, Shan and Lin Xiaoyi. 2020. "China to Build Historic Yarlung Zangbo River Hydropower project in Tibet." *Global Times*, November 29. www.globaltimes.cn/content/1208405.shtml.

Karmakar, Rahul. 2019. "After Years of Hydro Push, Arunachal Begins Scrapping Dam Projects." *The Hindu*, September 15. www.thehindu.com/news/national/other-states/after-years-of-hydro-push-arunachal-begins-scrapping-dam-projects/article29422880.ece.

Kashyap, Samudra Gupta. 2017. "Arunachal Tribals Oppose 10,000-MW Hydro-Electric Dam on Siang River." *Indian Express*, October 18. https://indianexpress.com/article/india/arunachal-tribals-oppose-10000-mw-hydro-electric-dam-on-siang-4895418/.

Klaver, Irene J. 2022. "Radical Water." In *Hydrohumanities Water Discourse and Environmental Futures*, edited by Kim De Wolff, Rina C. Faletti and Ignacio López-Calvo, 64–89. Berkeley: University of California Press.

Mimi, Raju. 2017. "The Dibang Multipurpose Project: Resistance of the Idu Mishmi." In *Water Conflicts in Northeast India*, edited by K.J. Joy et al., 218–230. Abingdon and New York: Routledge.

Modak, Sayanangshu. 2020. "Spotlight on Planet's Largest Hydropower Project by China on Yarlung/Brahmaputra." *Observer Research Foundation Online*, December 12. www.orfonline.org/expert-speak/spotlight-on-planets-largest-hydropower-project-by-china-on-yarlungbrahmaputra/.

Mohammed, Khaled et al. 2017. "Extreme Flows and Water Availability of the Brahmaputra River under 1.5 and 2°C Global Warming Scenario." *Climactic Change* 145: 159–175.

Ozguc, Umut and Andrew Burridge. 2023. "More-than-Human Borders: A New Research Agenda for Posthuman Conversations in Border Studies." *Geopolitics* 28(2): 471–489.

Paerregaard, Karsten. 2020. "Climing the Andes: Vertical Complementarity, Trans-human Reciprocity, and Climate Change in the Peruvian Highlands." In *Storying Multipolar Climes of the Himalaya, Andes, and Artic: Anthropocenic Climate and Shapeshifting Lifeworlds*, edited by Dan Smyer Yü and Jelle J.P. Wouters, 52–68. Abingdon and New York: Routledge.

Ramesh, Jairam. 2015. *Green Signals: Ecology, Growth, and Democracy in India*. Oxford: Oxford University Press.

Saikia, Arupjoyti. 2019. *The Unquiet River: A Biography of the Brahmaputra*. Oxford: Oxford University Press.

Smyer Yü, Dan. 2021. "Situating Environmental Humanities in the New Himalayas." In *Environmental Humanities in the New Himalayas: Symbiotic Indigeneity, Commons, Sustainability*, edited by Dan Smyer Yü and Erik de Maaker, 1–24. Abingdon and New York: Routledge.

Thomas, Kimberley Anh. 2017. "The River-Border Complex: A Border-Integrated Approach to Transboundary River Governance Illustrated by the Ganges River and Indo-Bangladeshi Border." *Water International* 42(1): 34–53.

Thomas, Kimberley Anh. 2021. "International Rivers as Border Infrastructures: En/Forcing Borders in South Asia." *Political Geography* 89: 102448.

van Schendel, Willem. 2002. "Stateless in South Asia: The Making of the India–Bangladesh Enclaves." *Journal of Asian Studies* 61(1): 115–147.

van Schendel, Willem. 2020. *A History of Bangladesh*. 2nd edition. Cambridge: Cambridge University Press.

Watts, Jonathan. 2010. "Chinese Engineers Propose World's Biggest Hydro-Electric Project in Tibet." *Guardian*, May 24. www.theguardian.com/environment/2010/may/24/chinese-hydroengineers-propose-tibet-dam.

Wouters, Jelle J.P. and Michael T. Heneise. 2022. "Highland Asia as a World Region: An Introduction." In *Routledge Handbook of Highland Asia*, edited by Jelle J.P. Wouters and Michael T. Heneise, 1–40. Abingdon and New York: Routledge.

Xinhua News. 2017. "Highway to Heaven, and to China's Most Isolated County." April 17. www.xinhuanet.com/english/2017-04/17/c_136214937.htm.

Xinhua News. 2021. "Highway through World's Deepest Canyon Completed in Tibet." May 16. www.xinhuanet.com/english/2021-05/16/c_139949882_3.htm.

Yingqing, Li and Hou Liqiang. 2019. "Bengal Tigers Found in Tibet, with Plenty of Prey." *China Daily*, August 8. www.chinadaily.com.cn/a/201908/08/WS5d4b646ea310cf3e355647c2.html.

10 Encountering Climate Change

Agential Mountains, Angry Deities, and Anthropocenic Clime in the Bhutan Highlands

Deki Yangzom and Jelle J.P. Wouters

Introduction

This chapter invites readers to the Bhutan highlands to reflect on the agential entanglements between humans and the mountains with which they have cohabited for thousands of years. This relationship is intimate, so that Bhutanese highlanders generally experience and conceptualize their own being, becoming, and belonging through enduring entanglements with specific mountains. As noted in Chapter 1, Himalayan mountains are variously the makers, recipients, and changers of climate. In the "New Himalayas" (Smyer Yü 2021), mountains are progressively anthropogenically affected, with manifold ramifications that span subsurface to atmospheric levels, micro-climatic to planetary spheres, and material to cultural–religious domains. To further apprehend the Anthropocenic Himalayas, this chapter coins and enlivens the notion of "mountain clime."

Continuing clime studies (Smyer Yü and Wouters 2023), our coinage of "mountain clime" highlights how climate change is terrestrially experienced and engaged in mountain environments. Also aligning with posthumanist currents, such as environmental humanities, new animisms and materialisms, and multispecies studies, our conceptualization of mountain clime emphasizes the integrality of more-than-human and more-than-biotic communities of life, along with their co-constitutive elemental entanglements. The Bhutan mountain context reaffirms the mutual embodiment of seeming opposites (seeming, that is, in globally dominant secular-scientific knowledge fields), such as the realms of the material and the immanent, of matter and meaning, of nature and culture, of object and subject, of abiotic and biotic, and so on. However, in conceptualizing a mountain clime we do not aim to perform presumptions of intellectual novelty in any true sense; rather, our chapter is best read as an affirmation of emplaced ontologies, epistemologies, and genealogies of knowledge and praxis in the Bhutan highlands.

Thinking with a mountain clime affirms how mountains have long been vital to highland human life, both for their life-enabling and sustaining glacio-hydrological and geo-ecological affordances, and because of their religious–cultural significance, in the sense that highland Bhutanese and their companion species

DOI: 10.4324/9781003484394-10

(yaks in particular but also horses and sheep) have long arranged their livelihoods, lifeworlds, sense of morality, and everyday ethics around mountains. Relatedly, and crucially, our coinage of mountain clime also affirms how local climing practices, in terms of how and what highlanders know, feel, and sense about weather, climate, geo-ecological relations, and changes therein, are refracted by their imagined and ritually mediated relationship with mountains. We use "imagined" and "imagination" here not in the sense of "imaginary," "fictitious," or "fabricated," but in reference to its anthropological and neuroscientific meaning of the varied, complex, and diversely culturally shaped connection between perception, experience, and reality.

As a form of integrally enmeshed "being–relating–knowing" in the Bhutan highlands, the situated, creative, and generative forces of the human–mountain imagination inherently involve nonsecular entities. These numinous beings identify as gods, deities, and earth spirits of Bon–Buddhist affiliation. Changes in weather and climate are locally received and related to as sentient perceptions and communications by these meta-human or more-than-human personages. Changes unfavorable to human flourishing reveal interruptions, transgressions, and breakdowns in the shared "norm-worlds" (see Banerjee and Wouters 2022) that are envisioned and enacted between humans, mountains, and nonsecular beings. Ritual action, broadly conceived, is generally summoned to bring these more-than-human and more-than-biotic norm-worlds back to equanimity.

While the presence and authority of nonsecular beings in Bhutan (and elsewhere) is a felt reality and inherent to local worlding practices, these understandings are notably, and notoriously, suppressed by ostensibly secular climate science. This suppression persists despite the observation that most people in the world do not think and experience climate change in primarily secular terms (Haberman 2021). Secular scientific approaches to climate change alone certainly do not carry significant convincing powers in the Bhutan highlands, where mountains generally emerge as dominant and agential actors. Across Bhutan, writes Karma Ura (2001), mountains "are the citadels of many great *lha* [deities], *nasdag* [lord of the soil/earth] and *zhidag* [lord of the settlement]." They are "immortal owners or landlords, while successive generations of [human] communities are ephemeral travelers passing through their territory." These "immortal owners," and their mountain citadels/bodies, co-emerge as agentive terrestrial life forces in the everyday business of ecological, social, and spiritual living. This makes mountains crucial members of Bhutanese society.

Our ethnographic elaborations from Bhutan's eastern and western highlands, where we individually followed the trails and tails of humans and yaks, revealed that mountains and humans are part of a never-ending conversation. This conversation gathers nonhuman and scientifically nonliving forces that assert themselves around, in, and through humans. While it is obvious that highlanders are materially, socially, and spiritually entangled and entwined with mountains, our fieldwork simultaneously revealed that it is obvious to highlanders that mountains pay attention to, hear, smell, and respond to human (in)actions. Changes in the physical landscape, including the retreat of glaciers (Cruikshank 2005; Gagné 2019) and the upward shift of the treeline, and in precipitation amounts, the velocity of winds, higher temperatures, and natural disasters are locally acknowledged as modes of mountains' communication and representation. A mountain clime, then, is a geologically, ecologically, spiritually, and socially shaped reality that has, in a sense, its own biography and "being" through emplaced forms of intersubjective communication, bonding, and interaction between its many beings, whether biotic, inorganic, material, numinous, or elemental (or a confluence of these).

Said otherwise, in a mountain clime, highlanders are both in the mountain and of it, act in relation to it and are acted upon, sense it and are sensed by it. Mountains, relatedly, are both "something" and "somebody," matters of substance and significance, geological fact and emplacement of religion. Ultimately, this mutualism of being and relations enacts a normative grammar that extends not only beyond humans to include other-than-humans but also beyond the biotic—that is, beyond what the classifications and boundaries of dominant scientific frameworks accept as alive and agential. By affirming the aliveness, awareness, and responsiveness of mountains, Bhutanese vernacular knowledge traditions of being–relating–knowing challenge and complement dominant scientific taxonomies of life and nonlife through epistemologies and spectrums of agency and intentionality that expand our understanding of sentience, perception, communication, and cognition. And this, as we shall illustrate below, has implications for how climate change is encountered.

Being–Relating–Knowing in a Bhutan Mountain Clime

Much of the course of Bhutan's history and society has been steered by human–mountain entanglements. This starts with the geologically, ecologically, and climatically shaped gradients and niches of the landscape, to which human and other-than-human life forms adapted. This centrality of mountains is underscored in Bhutan by the creative interplay between religion, political economy, meaning, and power. The Buddhist saint Padmasambhava, known to his devotees throughout the Himalayas as Guru Rinpoche, who arrived in the region in the eighth century CE, is "without any doubt the most important and universal of all historical and religious figures in Bhutan" (Phuntsho 2013, 84). He is credited with taming a pantheon of powerful, initially often human-averse autochthonous deities, many of whom resided in the mountains (but also in bodies of water, on cliffs, in caves, and in or around other landscape markers). He bound these gods and goddesses under oath to serve and protect the Buddha dharma (Aris 1979; Phuntsho 2013), but also affirmed their status and rights as first-settlers in the landscape. In doing so, he secured permission for humans to settle, traverse, cultivate, and flourish in Bhutan, on condition that they paid regular ritual homage and everyday acquiescence to the "immortal owners" (Ura 2001). This conviction that deities are emplaced, even if unseen, territorial sovereigns remains a dominant narrative across Bhutan, bridging Bon and Buddhism, though at times in complex and contested ways (Tashi 2023), and became integral to the country's environmental and development policies (Allison 2019). In their entanglements with humans, Bhutan's mountains are reminiscent of Andean "Earth beings" (a translation of the Quechua word "*tirakuna*"; see de la Cadena 2015). These are sentient entities that do not merely inhabit but *are* mountains, rivers, lagoons, and other visible landscape markers. They exist in a social world of *ayllu*—the relationship between themselves and the local human population. However, this notion needs adapting to the Bhutanese context, where "Mountains may or may not be the residence of *yul lha*, the deity of the territory, or may be the deity himself or herself" (Pommaret 2004, 41). *Yul* variously translates as territory, home, village, or native place, while *lha* is a deity or god. So *yul lha* evokes a terrestrial and territorial deity, or an earth god/goddess, and simultaneously symbolizes and delimits a territory over which that deity presides as an unseen sovereign. As noted, *yul lha*, as well as being cognate categories of deities, are often associated with mountains, which can be their embodiment (*brtan*) as well as their abode (*gnas*), which is a complementary understanding of sentient and sacred space compared to Andean Earth beings.

Yul lha mountains, whether as *brtan* or *gnas*, as Toni Huber (1999, 23) elaborates with regard to Tibet, are "an important ritual basis for the constitution and vitality of local communities." This is a ritual basis that we also understand as a human acknowledgment of mountains' life-giving geo-ecological, hydraulic, and climatic affordances. Across the Bon–Buddhist Himalayas, mountains are the "most venerated and culturally significant feature of the landscape," and, in that capacity, "function as one significant form of organizing principle" (Huber 1999, 21) of space, society, and history. To imagine and engage with mountains as a clime affirms their "organizing principle" in cultural–religious terms but also expands this to include a more-than-human "life principle," with mountains as a precondition of, and immanent in, the material condition of highland life. As Deep Time geological creations and climate-makers, mountains antecedently provide the material and meteorological context on which highland humans and other-than-humans live and die. In a mountain clime, their affordances as enablers, sustainers, and shapers of life are acknowledged by the animating principles of religion, culture, and identity, and by the varied ways in which they are enlivened, ensouled, and personalized in material–metaphysical schema. Thus envisioned and enacted as living materialities, mountains enmesh humans and other-than-humans to form lively communities of life and shared norm-worlds in which humans generally play a dependent part (Wouters 2023).

These shared norm-worlds avow the palpable connections between mountains, including their residential/embodied deities, and human lives. In their everyday enactments, they encompass ritual performances, everyday propitiations and prayers, restrictions on human activities (e.g., there should be no indiscriminate rock-breaking or tree-felling), and prohibitions on causing *drib*—a foundational and relational principle that structures ethics, values, and moral reasoning in the Bhutan highlands, as well as textually in Buddhism. Closely associated with *dik*, or sin, *drib* variously signifies pollution, contamination, or defilement that can be bodily, spiritual, social, moral, or cognitive and which—in both Bon and Tibetan Buddhism—impedes and complicates productive beyond-human relations (i. e., with deities) as well as individual spiritual and karmic progress. The currently unfolding Anthropocenic mountain clime, in terms of mountains' altering affordances and fracturing norm-worlds, is discussed locally in terms of an accumulative overflow of *drib* and *dik*, so that the aforementioned organizing, life, and relational principles of mountains are reshuffled.

Bhutanese mountains' experienced animacy and role within an embracing cosmos connect to what Dan Smyer Yü (2020) innovatively terms the "affective consciousness of the earth," "metaphoric animism," and the "planetary animist sphere." These phrases affectively apprehend how—in diverse conceptions and contexts, from Native American and Australian Aboriginal to those of the Himalayas and the Tibetan Plateau—the interior life force of the earth, understood in spiritual, psychic, and sentient terms, can "incarnate in and interlace with different exterior bodies" (Smyer Yü 2020, 273) whether of organic beings or earth materialities. In terms of geo-physiology, humans cultivate an intimate capacity to understand and engage with the interactions and relations that conjoin their eco-climatic niches, or climes, including changes therein. Tim Ingold (2000) employs the phrase "sentient ecology" not only to emphasize the overall livingness of the environment but also to indicate forms of knowledge that draw from emplaced "intuition," based in feelings, sensitivities, skills, and orientations that emerge from conducting one's life in a particular eco-climatic niche. While intuition is often contrasted with the logic/products of the rational intellect, including climate science, Ingold (2000, 25) convincingly argues that no meaningful intelligence is completely detached from the condition of life in the actual world, and that intuition is a "necessary foundation for any system of science or ethics," including those of Euro-American genesis and evolution. He concludes:

> Intuitive understanding, in short, is not contrary to science or ethics, nor does it appeal to instinct rather than reason, or to supposedly "hard-wired" imperatives of human nature. On the contrary, it rests in perceptual skills that emerge, from each and every being, through a process of development in a historically specific environment.

Our coinage of "mountain clime" translates and transmits the incarnation of a living and lively earth into agential material bodies such as mountains, as well as the emplaced "intuition" and other forms of premonitory knowledge that so often inform the Bhutanese highlanders' understanding of their surroundings. It irrigates a "being–relating–knowing" that shapes emplaced practices, morality, and ethics of "ritual geology"—a term we borrow and adapt from Robyn D'Avignon (2022) to relate how geological constructions are affectively experienced and ritually addressed. It is situationally productive to think of a mountain clime in the Anthropocene context because it relates a transmission of ancient Bhutanese knowledge of the agentive capacities of mountains, as enlivened by resident deities.

In summary, thinking in terms of a mountain clime highlights how, in Bhutan and across the Bon–Buddhist Himalayas, mountains have long co-designed human societies, just as humans have co-designed the materiality, meaning, memory, and value of mountains. Affirming this mutualism of being and relation, our coinage of "mountain clime" coalesces the mountains' materiality and meaning, the profound material and moral forces that co-shape an agential world of which humans are a part. It thus transmits mountains as both a materiality that offers real (and changing) physical affordances of highland life and as willful beings, in the sense that they are imbued with substantial common cultural understanding and, in their manifestation as *yul lha*, exert agency in social worlds that are currently reorienting.

Meeting Ama Jomo

In the remainder of this chapter, we enliven our coinage of "mountain clime" through two ethnographic narratives that emerge from Bhutan's western and eastern highlands. These ostensibly disparate settings merge on a cosmic plane, as both pivot around the same powerful, terrestrial goddess—Ama Jomo. Moreover, both places are closely associated with yak-herders, which is also true of Ama Jomo herself, who is sometimes referred to as the "god[dess] of yaks." While Ama Jomo is far from the only terrestrial deity in these highlands and mountains, she certainly has a dominating presence.

The first narrative emerges from Wouters' ongoing research in the western highlands, and specifically from a conversation about local climate change with Sonam Dorji, an elderly yak-herder. The second is a "walking ethnography" compiled during Yangzom's clim(b)e to the palace of Ama Jomo in eastern Bhutan. Extending from our ethnographic insights, the final section reflects on how climate change is imagined and engaged in a mountain clime, which is experienced as a breakdown in the more-than-human and more-than-biotic worlds that are arrayed around mountains. Rather than seeing this as a purely local and alternative perspective on climate change, we emphasize the need for a productive exchange between global, scientific discourses about climate change and emplaced, more-than-human imagination and experience to arrive at more capacious, inclusive, and deeper forms of knowing.

Like most *yul lha*, Ama Jomo arose out of Bon beliefs and eventually found a place in contemporary Buddhism. She arrived in Bhutan from a place named Tshona in Tibet (which she is still rumored to visit occasionally), where she freed a community of bonded laborers from a tyrannical king by advising them to decapitate the ruler rather than behead a local mountain, as the king had ordered them to do in order to increase the amount of sunlight in his palace. After the laborers killed the king, Ama Jomo helped them escape by guiding them to eastern Bhutan. To this day, the people of the region—the Merakpas—identify as the escapees' descendants. Indeed, Yangzom's fieldwork reveals that Ama Jomo remains central to the existence, identity, and sustenance of this community. It must be noted, however, that this narrative is not unique to the Merakpas; it also figures prominently, albeit with some alterations, further east in Tawang, Arunachal Pradesh (Gohain 2020, 117), and in the highlands of western Bhutan.

"The Snow Has Disappeared!" Or: The Poison of the Heat

"*Gangri muti zhaynu*" translates as "the mountain is wearing lustrous pearls." It was how highlanders used to describe the glaciers and snow that adorned Jomolhari. But not anymore. "*Grangri khaw zhu nu* [the snow has disappeared from the mountain]," yak-herder Sonam Dorji sighs, pointing toward Jomolhari, which rises in the distance.

Jomolhari—or Jomo's divine mountain (Pommaret 1994)—houses a citadel in which Ama Jomo (periodically) resides. This can be reached by ritually crossing a geologically shaped door (*neygoh*) to a sacred place. Standing over 7000 meters tall, Jomolhari is a space of ongoing material, cultural, and spiritual exchange that sprawls across the anthropogenic border between Bhutan's western highlands and Tibet. The mountain overshadows the extensive grasslands that extend from her base (see Figure 10.1). These plains, which are irrigated by (increasingly shallow) streams that originate in Jomolhari's glaciers and snow, nourish Sonam Dorji's and other highlanders' yaks, whose presence in this region enables human life (Wouters 2021).

We are in Soe Village, which consists of a sprinkling of houses, government offices, and a single shop. A two-day walk from the Paro Valley, where the road ends, it is where Sonam Dorji and other herders reside when they are not co-migrating with their yaks and horses. In recent years, many highlanders have also procured land and built houses in the Paro Valley with wealth generated from the collection and sale of "caterpillar fungus" (*Ophiocordyceps sinensis*), which is in high demand for its medicinal properties. As has been variously noted, in Bhutan and elsewhere in the Himalayas, this new fungus economy is transforming highland livelihoods and lifeworlds (Choki 2021; Winkler 2010).

My conversation with Sonam Dorji is primarily about the changing climate. It is a topic about which he has much to say, and his prime reference point is Jomolhari. "When I was his age [pointing to his grandson, who is playing nearby], Jomolhari was covered with snow, thick layers of it, certainly in winter," he says as we both stare at the mountain. "Now the snow is less. See the black spots everywhere? It looks like a disease. The mountain is sick."

Sonam Dorji is referring to Jomolhari's changing chromaticity—from snow-white to the black of rocks stripped of their cover. These rocks have been snow-clad since at least the Late Pleistocene. Until recently, that is. He explains further: "Earlier, Jomolhari was always white, but now in the third or fourth month [of the Bhutanese calendar] the heat melts it [the snow] away."

Figure 10.1 View of Jomolhari in Bhutan's western highlands
Source: Photo by Pema Choden

Not only has the appearance of Jomolhari changed. So has its sound: "The ice now cracks when we walk over it. The sound of it is scary. Earlier, ice didn't make a sound." Earth sciences and climatology explain that retreating glaciers and less snow lead to a lowering of albedo—the amount of solar radiation that is reflected from the earth back into space. Black surfaces, such as the now-exposed rocks of Jomolhari, absorb heat, whereas lighter surfaces, such as those covered with snow, reflect it. Changes in albedo have both local and global effects on temperature. In the Jomolhari area, the decreasing albedo multiplies the effect of the already rising average temperatures. Sonam Dorji describes the new heat he experiences as *dhu*—poison. This reveals itself in various ways. Winters are becoming like summers, he says. There is no longer much difference between temperatures in the highlands and in the Paro

Valley. And whereas, in the past, the ground was so hard that they needed to burn the end of a post to force it into the ground, today the post goes in with no more than a little pressure.

"What is causing this *dhu*?" I ask, expecting a lament about global emissions, perhaps with particular blame cast on the materially saturated West and the pollution that comes to Bhutan from India or China, depending on how the wind blows. After all, Bhutan is carbon negative, in the sense that its extensive forest cover absorbs more greenhouse gas than the country produces.

Surprisingly, though, Sonam Dorji invokes causal connections that are locally grounded, laying the blame on the highlanders themselves. "If we don't follow our rituals and culture properly, how can we expect Jomolhari to protect and provide for us?" This is a recurring theme in our conversations. On several earlier occasions, Sonam Dorji has expressed disapproval of the younger generations' apathetic attitude towards rituals and tradition. Several other—mostly elderly—yak-herders voice similar concerns about changing lifestyles. In particular, they bemoan young herders' obsession with "the material," in terms of generating wealth and accumulating everyday comforts, and neglect of their roles and obligations in extensive and complex ritual and social norm-worlds that integrally affirm and engage Ama Jomo and a range of other terrestrial deities.

The expectations, norms, and taboos associated with Ama Jomo are familiar to all highlanders. Besides regular ritual homage, propitiations, and offerings, the situated more-than-human, more-than-biotic norm-world prohibits the burning of meat and chilies (the smell of which Ama Jomo dislikes), discarding waste, making loud noises near sacred abodes, indiscriminately cutting down trees and digging the soil, and generating *drib*. Straightforward as these rules may seem, Sonam Dorji and other elders are witnessing ever more transgressions. He says: "There is an old saying that gods, devils, and humans have the same behavior. When we are hurt by people, we change our behavior towards them. The same applies to Ama Jomo. She doesn't approve of our actions. They make her angry." This anger is manifested in more disease (among both humans and yaks), vanishing snow, and unusually strong winds. Affirming Jomolhari as a climate-maker, Sonam Dorji, like most of the yak-herders in Soe, ascribes the changing weather to Ama Jomo communicating her displeasure to humans. Additionally, wild animals are becoming unusually bold. "When we don't perform the *norchoe* ritual for the well-being of the yaks, snow leopards and bears visit our village," he says. "They kill our young yaks. Sometimes they even knock on our doors. Seeing such threats, I feel it must be the wrath of Ama Jomo." If these signs are not already sufficient evidence, local *pawo*s (shamans) confirm, through divination, that Ama Jomo is indeed angry. They prescribe ritual remedies that the herders carry out, albeit to no effect (so far). Sonam Dorji and other highlanders observe this diminishing efficacy of ritual with apprehension, interpreting it as a sign of the unprecedented extent of Ama Jomo's displeasure.

I probe further into the reasons for Ama Jomo's anger. Sonam Dorji highlights the arrival of electricity, development programs and projects, and especially the trade in caterpillar fungus, which he feels have changed not only the material standards of highland living but also local mindsets, resulting in lifestyles that are far removed from traditional yak-herding. In drawing this link between the new fungus economy and cosmic imbalance, he corroborates Kinley Choki's (2021) account of recent developments in Lingzhi Village (a climb–descent–climb from Soe), where the trade is generating "unimaginable wealth" and transforming the inhabitants' pastoral lifestyle and cosmovision. For instance, there has been a steep decline in community and household rituals and a steep increase in material possessions, including large, permanent houses. Elderly villagers (Lingzhips) explicitly accuse the fungus economy of causing the deities' displeasure, which they experience as erratic weather, poor health among humans and animals, decreases in the quantity and quality of pasture, and a palpable increase in natural disasters. "[T]hey are well aware of their responsibility towards the local deities as much as the deities are of their responsibilities towards the Lingzhips," but as this mutual responsibility continues to unravel, "the angered deities are causing changes in the climate and landscape that are not benefiting the Lingzhips in any way" (Choki 2021, 161).

Returning to our conversation, Sonam Dorji also highlights the increasing number of outsiders in the highlands, including tourists, soldiers, and government officials. They enter Ama Jomo's *yul* (territory) without always being aware of her expectations, restrictions, and taboos. This means they more readily transgress the deity's proscriptions, although it is the permanent residents—the herders—who invariably suffer the consequences. "Outsiders often burn waste, including meat. This is not acceptable to Ama Jomo," Sonam Dorji reports. He is most critical of the guides who accompany tourists into the region. "When this happens, Ama Jomo changes the weather. She won't give us what we want."

An old article in *Himal* (Phuntsho 1992) corroborates Sonam Dorji's lament about tourists, angry deities, and changing weather. Written in the early 1990s, it discusses local experiences and reactions to the opening of Bhutan's mountains to mountaineering expeditions over the preceding decade. For instance, a representative of Lingzhi complains about the arrival of the mountaineers: "Chomolhari [Jomolhari] is the residence of Chomo [Jomo], the deity who watches over our herds. It is a monastery where we offer our prayers." Another peak that attracted mountaineers was Jitchu Drake, next to Jomolhari. The article relates how

> the yak herders were aghast when they learnt that the climbers meant to trample atop their revered peak [as] defilement of the summit meant that the wrath of gods would manifest itself in bad weather and the spread of diseases … [Thereafter,] the weather became unusually hostile, with hail and wind while the sun still shone.

A *gup* (head of a *gewog*, or cluster of villages) from the Jomolhari region made his objections clear when speaking to the reporter: "I always pray to the local deities for snow when you outsiders come to our valley, so you will go away. You use all our firewood and show little respect for our tradition."

Five years earlier, a special commission had studied the relationship between tourism, the abodes of deities, and religion. Its secretary reported to the national assembly in 1987:

> If firm and timely measures are not taken to protect the aura of sanctity that still pervades most of our sacred places of worship, not only will our own reverence and faith be undermined but the belief and faith of our own children in our religion and culture will be placed in jeopardy.
> (Cited in Phutsho 1992)

Some restrictions were duly implemented, including the closure of Jomolhari to mountaineers. However, the yak-herders continued to link weather changes to climbing expeditions, resulting in an extension of the ban to encompass every peak over 6000 meters in the mid-1990s. Finally, in 2003, all mountaineering was banned throughout the country. Unfortunately, though, this has done little to stop government officials and soldiers from transgressing within shared norm-worlds, as Sonam Dorji is quick to point out.

Of all the consequent changes in climate and weather he observes, what worries him most is the disappearance of snow:

> If the snow melts, the water dries up. Before, during the summer, the [surface] water was big; and when it was winter, the water was smaller. But now the water doesn't get big even in the summer. Once all the snow has disappeared, all of Ama Jomo's *jinlab* [blessings] will disappear too.

There are both ecological and spiritual dimensions to this assessment of the unfolding water crisis. In terms of ecological services, Jomolhari is the source of Paro Chu (Paro River), which flows from the south side of the mountain, and Ama Chu, which flows from the north side into Tibet. Both provide life-giving, life-nourishing water to humans and nonhumans alike. However, the retreat of the mountain's glaciers and a decline in snowfall have resulted in lower annual snow-melt and therefore less water on the plains. From a spiritual perspective, the increased scarcity of water is Ama Jomo's response to the recent breakdown of mutualistic relations in the highlands.

According to Sonam Dorji, the unprecedented heat and droughts of recent years are not altogether surprising, as both were predicted long ago:

> My grandparents used to say that when the world is going to end, we will experience the immense heat from a combination of seven suns. When this happens, the world and all its humans and animals will experience such unusual heat that not a single species of animal or plant can survive.

The water and soil will all get dried up and no food will be available. Conflicts will increase between the people, animals that live will find it difficult. As a teen, I didn't bother much when they said such things, but now I realize what they meant. Jomolhari turning black means that life is in peril. Our rituals will no longer be answered. There will be more disease, droughts, and war. If this continues, we will reach *zamling kabenob* [the end of the world].

Clim(b)ing Ama Jomo

We now travel from to the eastern highlands of Bhutan, where Yangzom clim(b)ed to Ama Jomo's *phodrang* (palace) in Merak. Climbing also becomes climing when the agency of the Deep Time geological landscape, climate, and weather is imagined and engaged through spiritual representations and ritual interaction. There are several similarities between this region and the western highlands. First, climate change in the mountain clime of Merak interlocks with the local herders' awareness of agential mountains enlivened by emplaced/embodied nonsecular beings. Second, the region's primary and most powerful deity is Ama Jomo. Third, herders in Merak frequently mention *namshey jurwa*, which literally translates as "weather change" but also evokes structural changes in mountain ecologies and experienced climate. Fourth, elderly herders ascribe these changes to the rapid expansion of the built environment (roads and buildings), coupled with the erosion of community traditions and ritual life, both of which are disrupting long-standing relationships between the Merakpas, the mountains, Ama Jomo, and other resident deities. For example, they believe that excavation equipment is a prime cause of climate change as its use angers the deities.

It is during my second period of fieldwork among the Merakpas, who traditionally are nomadic yak-herders, that I decide to clime the steep climb to Ama Jomo's *phodrang*. It is early spring, but the weather is still wintry and the wind piercing when I share my intention with the local herders. I ask whether any of them would be willing to accompany me. They all express reluctance and try to dissuade me. The weather is too harsh, they say, and altitude sickness is a real possibility, especially for a city-dweller like me. It is only later that I learn about another reason for their lack of enthusiasm. No herder has visited Ama Jomo's abode so far this year, and the first visit will have to be flawless to garner her favor for the months ahead. Moreover, while Ama Jomo relishes visits to her abode, she prefers male callers, all of whom she regards as her human husbands. (Merakpas often talk about their husband status in relation to Ama Jomo, usually light-heartedly.) By contrast, the deity sometimes sees women as competitors, so the possibility of my arrival upsetting her was another reason for the herders' reluctance to escort me to the palace. Nevertheless, a few evenings later, as we sit around the fire, Singye, whom I have come to know well, offers to accompany me, probably concerned that I will not be safe alone in the mountains.

The next morning we meet Singye's cousin, Kinzang, who has decided to join us on the trek. The sky is gray and the mountain is shrouded in mist. Light rain is falling. Singye hesitates and remarks that the weather is unfavorable. He asks whether I have followed his instructions. I confirm that I have not eaten pork, chicken, egg, garlic, or onion. Ama Jomo does not like the smell of these foods and knows immediately if any of her visitors have consumed them. Singye seems satisfied with my response and we set off. However, as we ascend, the drizzle turns into a proper rain shower. Singye stops suddenly and asks for my backpack, which he opens. Rummaging through it, he finds two packets of "Wai Wai Noodles" I had brought as a snack. After scrutinizing the contents, he announces, "These have onion and garlic in them." The disappointment is thick in his voice. He places the packets on the side of the path. "We can pick them up on our return." Soon after, the rain ceases and the sky clears. "I knew that something was wrong," Singye says with a grin. "Ama Jomo speaks to us by changing the weather. We experience this all the time. Your noodles would have polluted her palace. The rain was her giving us advance warning and a chance to correct ourselves."

Every Merakpa knows that Ama Jomo communicates through weather events, so interpreting meteorological changes is an everyday activity. Just a few days earlier, a yak-herder had told me about an incident with an excavator they had hired to build a road to the community school:

> The weather was pleasant that day, but as soon as the engine of the JCB was switched on, dark clouds arrived and hail fell down. The weather did not allow us to do the work. We realized that this was Ama Jomo speaking to us. We consulted a *tsib* [ritual master], who informed us that proper rituals had not been conducted and that no approvals had been obtained from Ama Jomo to carve the landscape. This was our mistake. Once the ritual was done, the weather cleared and the JCB was allowed to do its work.

Singye, Kinzang, and I arrive at a place called Serkem Pangthang ("land to make an offering"). This is the entrance that leads to Ama Jomo's palace. Singye takes out a bottle of *ara* (fermented drink) to prepare a *serkem* (alcohol) offering. As he sprinkles the alcohol in all directions, he closes his eyes and chants:

> *due chi gi lo lu*
> *Jomo Phodrang jewa joda*
> *key ngen meyba,*
> *nam shey zang tey chab ze bar sho,*
> *tshog lu zhey bar sho.*
>
> For the first time, this year,
> As we enter the place of your sacred abode,
> Protect us from peril,
> May you bless us with favorable weather throughout our journey,
> And accept our genuine offering.

He emphasizes that we are the first people to pay homage to Ama Jomo this year, so we must perform all the offerings, propitiations, and prayers carefully and correctly. This requires constant spiritual and ritual interaction with the Deep Time geological structures over and through which we maneuver. For example, when we reach a small stream, without warning, Kinzang splashes water over me. Surprised, I jump back to prevent the water from touching me. Singye admonishes me and explains that this water spiritually cleanses our bodies, something Ama Jomo expects of us.

We begin the steep, arduous ascent to Ama Jomo's palace. Two hours later, we arrive at the first summit. Singye looks for dried twigs in order to make a fire for a *bsang* (incense smoke) offering. In addition to being a purification rite, this will communicate to Ama Jomo and the other deities who are emplaced/embodied in the landscape that we are nearing. Singye burns the incense while reciting the deities' names—Ama Jomo, Lama Jaripa, Lama Khijay, Lama Kencho (*gnasbdag*), Mukuling (*btsan*), Dzambala, Gara Wangchuk, Lopen Tshotshong, and Lopen Serphu (*kyi lha*)—and asks them to accept our offering. All these deities arrived here with Ama Jomo. Mukuling has since relocated his palace, but he must still be acknowledged.

Continuing our climb, we arrive at Ta Juk Thang ("horse-racing ground"). This is where Merakpas hold an annual horse race during the Jomo Kora—a day-long ritual conducted on the fifteenth day of the eighth lunar month, when grand food offerings are made to Ama Jomo. The race commemorates Ama Jomo's journey from Tshona in Tibet to the Merak highlands, which she accomplished on a horse. Next, we cross Chum Dur Dur ("sound of invisible water"). This is where Ama Jomo searched for water to cook food. While the sound of water was clearly audible, it was nowhere to be seen. Ama Jomo therefore created a *drup chu* (holy water) spring nearby. Singye tells me that one visitor tried to find the invisible water by lowering her *zaa phob* (a precious and expensive wooden cup) between the rocks, but a *chu gi lha* (water deity) snatched it away.

After another three hours of climbing, Ama Jomo's abode finally comes into view for the first time. Nevertheless, the journey still seems endless. Singye remarks that the sanctity of her abode is such that, although we can now see it from afar, it remains hard to reach. He explains that the mountain was prophesied as the abode of Ama Jomo by Lama Jaripa, who accompanied her on the journey from Tshona to Merak. We cross several lakes that are nourished by glacial- and snow-melt, then the snowline. A thick veil of fog descends and the temperature suddenly drops. Observing the change in the weather and interpreting it as another message from Ama Jomo, Singye and Kinzang prepare a food offering. Singye grinds some biscuits on a *rangthang*—a large flat rock—and chants:

> *wur war wur war,*
> *Ama Jomo ghi ney sa lu,*
> *tshog chay ba phil tey kewa zang,*
> *zu kham zang tey sang phel ghi lo lu ya phil bey moenlam yoe*

[Verbalizing the sound of the flattened rock]
We are here at Ama Jomo's sacred site
It's a privilege to make this food offering
May we be in good health to make such a food offering next year too.

Kinzang and I then take turns grinding the biscuits. The fog continues to thicken and we are freezing. Breathing is becoming more difficult, but we continue our journey until we reach Gogolama ("the ultimate mountain pass"), which is another place of ritual interaction before arriving at the *phodrang* There is a chorten (Buddhist shrine) here, and Singye tells me that new visitors must make a *lui choepa* (song offering) to Ama Jomo. At first, I suspect he is teasing me, but he insists that my *lui choepa* will clear the fog

Figure 10.2 Crossing the snowline en route to the sacred citadel of Ama Jomo
Source: Photo by Deki Yangzom

because Ama Jomo enjoys listening to human songs. Reluctantly, I oblige. Once again, Singye's intimate knowledge of Ama Jomo's likes and dislikes proves infallible. The fog clears and the path that leads to the deity's abode becomes visible.

We walk over a steep ridge. Ama Jomo's *phodrang* is clearly visible now. The mountain stands guard over it. We arrive at a small *lhakhang* (temple) on a mound next to a dark lake. At that moment, the fog descends again and obscures our view of the mountain. Singye heads into the *lhakhang*—which contains a pictorial representation of Ama Jomo—and makes an offering by lighting butter lamps and incense. As expected, the fog clears again. Interpreting this as a "welcome sign," Singye is visibly elated. We prepare a further offering of milk and food for Ama Jomo and Dzambala (the god of wealth), who resides inside the lake.

Figure 10.3 Lhakhang on the shore of the lake below Ama Jomo's palace
Source: Photo by Deki Yangzom

Figure 10.4 Ama Jomo's lake and mountain abode
Source: Photo by Deki Yangzom

Shortly thereafter, Singye and Kinzang start to walk up to Ama Jomo's palace, which requires a steep climb from the lake. Singye tells me that menstruation makes women a source of *drib*, so only men are allowed inside. He then indicates a point I must not cross. Were I to do so, he adds, Ama Jomo would certainly react violently, making our return journey difficult, or maybe even impossible. I watch Singye and Kinzang complete the climb and once again wonder why a female deity does not allow a fellow female to visit her.

Following their return to the lake, Singye explains that it used to freeze over, but that is no longer the case. Kinzang adds:

> Look [pointing up at the *phodrang*], when you go up, there is a *makpoen phodrang* [warrior's palace] on each side of Ama Jomo's abode. The

> *makpoen* are the guards of her abode. Not long ago, the left *makpoen phodrang* eroded when the soil became loose due to heavy rainfall.

In the face of what the herders call *namshey jurwa* (here referring to anthropogenic climate change), even Ama Jomo and her *makpoen* are vulnerable to the *jungwa zhi*, the four elements: earth, water, fire, and wind. It is surely only a matter of time before the right *makpoen phodrang* suffers the same fate.

Prior to today's clim(b)ing, I met a lama who told me:

> I have spent three summer retreats at Ama Jomo's *phodrang*. What I witnessed was unearthly and you might not believe me when I say it now. It is like the mountain, the lake, and the landscape have sense and perception of what is forthcoming. They know it. During one of my meditations, the weather was perfectly fine but suddenly, seemingly out of nowhere, the sky turned dark with a gust of wind that persisted. I wondered what was happening, but [then] some visitors arrived. The changing weather was an indication of human arrival. On another occasion, the rain poured down, and this time I predicted that some people might be coming. I turned out to be right. Around Ama Jomo's palace, a sudden change in the weather is a prediction by the lake and the mountain.

Today's experiences seem to corroborate the lama's account.

Later, we climb a ridge to get a clearer view of a pair of twin lakes. Beyond them is a third lake where Ama Jomo bathed on her journey. Singye points towards a distant valley: "That place is called Tshogabu [literally, 'near a lake']. Do you see those huts? The herders bring their yaks there during warmer months." I ask if we might encounter some on the return journey. Singye replies: "We might meet a few herders, but no yaks." He explains that it used to be impossible to reach Tshogabu at this time of year because of the snow and cold, but now herders come and make arrangements long before they bring their yaks. He sighs, "It is *namshey jurwa*," and starts walking.

Encountering Climate Change

In the context of Andean waterways, Mattias Borg Rasmussen (2015; 2016) inverts dominant discussions about climate change adaptation—with peoples everywhere seeking to lower the risks posed by its consequences—by asking instead how climate change adapts to local life. As a planetary force, climate change encounters and adapts to various material, social, political, historical, spiritual, and affective worlds. In Bhutan, we witness the refracting (a term we think is more accurate than "adaptation") of climate change through the prism of the agential entanglements between mountains and humans, as the testimonies and actions of Sonam Dorji, Singye, and Kinzang illustrate. Simultaneously, this interpretive refracting does not fully fit the pattern of historical relations between mountains and humans, in the sense that climate change, in

Figure 10.5 Twin lakes in the vicinity of Ama Jomo's *phodrang*
Source: Photo by Deki Yangzom

its unprecedented rates and forms, is an encounter that is not just translated into but also challenges and transforms the relations of joint-becoming in a mountain clime. As it refracts, clime change therefore also probes and generates in its varying capacities to act upon and reorient the imagination. In Bhutan's mountain clime, this involves the orientation of climate change as a ritual and religious problem, and as a source of moral and ethical reflection and evaluation among members of contemporary society. Thus, while climate change has the capacity everywhere to work transformations in and of itself, these transformations can nowhere be understood apart from the particular historical and cultural circumstances that shape and condition their effects.

The ways in which climate change is refracted by Sonam Dorji, Singye, Kinzang, and by highlanders more generally, is seen in their Bon–Buddhist heritage and emplaced more-than-human and more-than-biotic norm-worlds that are grounded in intergenerational experiences of highland life. Their climing practices, in the sense of socially and ritually relating, and intuitively sensing, their geo-ecological surroundings, affirm mountains—in their material, spiritual, sentient, and affective relationships with humans—as agentive actors in the co-production of human subjectivity, meaning, and experience. In turn, the changing climate and environment, as locally encountered, are experientially related to breaches and breakdowns in the ritual, social, and sacrificial exchange between humans, mountains, and nonsecular *yul lha*, so that the previous mutualism of being and the relations between humans and mountains are disrupted.

In what anthropologists sometimes call "cosmic refusal" (Dowdy 2022), Ama Jomo, along with other *yul lha*, has no motive to conform to the currently exaggerated anthropogenic activities that attempt to change highland material and normative arrangements. As in neighboring Sikkim, where new ecological precarities and natural disasters are experienced and explained in an idiom of anxiety over deities enraged by anthropogenic transformations (Gergan 2017), the various narratives in this chapter reveal highlanders' experiences of deities' anger and so offer an indigenous critique of the modern scientific and secular tendencies of dominant Anthropocene and climate change knowledge, in which there is generally no place for spiritual–cosmological indeterminacy. In their acts and reasoning, highlanders center the agentive thrust of climate change in more-than-biotic entanglements, animacy, and cosmopolitical processes. Sonam Dorji, Singye, and Kinzang enact and envision mountains simultaneously as geo-ecological constructions and profoundly social, at once material and moral, and always sacred, and therefore as the recipients of human ritual action. Most of these rituals, including propitiations and prayers, are pragmatic, in the sense that they seek the reproduction of the mundane (see Huber 2020), especially the flourishing of humans and yaks. As such, ritual life in the highlands significantly focuses on (re)assuring fruitful mountain–human cohabitation, rather than, say, attaining redemption or the transcendence of this world. This focus on the mundane also reflects the character of most *yul lha* with whom highlanders co-dwell because these are themselves "worldly gods" (Tashi 2023) that have risen out of Bon traditions, found a place in contemporary dominant Buddhism, and are caught up with humans in the cycle of all existence, life, and matter. This also means that Ama Jomo and other worldly gods will attune to the emergent Anthropocenic clime, rather than escape from it.

In a very different context—that of gold mining in West Africa—Robyn D'Avignon (2022, 5) coins the phrase "ritual geology" to describe "a set of practices, prohibitions, and cosmological engagements with the earth that are widely shared and cultivated across a regional geological formation." In Senegal, but with parallels across the Sahel and savanna regions, the indigenous mining tradition—as opposed to industrial mining—entails ritual, cosmological, and cognitive engagements with the geological world. For one thing, gold is the property of earth spirits and mining it requires a sacrificial exchange relationship. For another, gold prospecting relates emplaced and embodied subterranean knowledge that is recorded and transmitted through bodily practice, indigenous tool-making, and storytelling embedded in landscape features and family genealogies. Thus, while geological processes are central to the creation of gold-bearing minerals, humans' attempts to access and extract it are characterized by a sacred and relational engagement with those processes and geological formations.

What applies to the quest for subterranean knowledge in West African goldfields extends to terrestrial, meteorological, and climate knowledge in the Bhutan highlands. Here, geological constructions are animated ritually and relationally, while pervasive cultural notions such as karma, causal connections, purity, sin, and pollution often co-emerge in relation to emplaced nonhuman beings, whether natural, numinous, or both. "Ritual geology," in the Bhutan highlands, further adjoins with what we might call "sacred meteorology," "moral clime," or "spiritual weather" to convey how weather events and patterns are locally linked to human behavior and ritual (in)action that affirm mountains as terrestrial weather- and climate-makers.

Yangzom's clim(b)ing with Singye and Kinzang to Ama Jomo's *phodrang* confirms how meteorology communicates the quality of human relationships with mountains and their nonsecular residents. This intersubjective logic also persists across Bhutan's boundaries. The most salient feature of Tibetan understandings of weather, Huber and Pedersen (1997, 577) posit, "is the experienced relationship between the vicissitudes of the physical climate and those of the prevailing 'moral climate' created by human activities." As such, the dynamics of the social world interact with and impact upon the weather, as co-created or directed by deities. Huber and Pedersen importantly affirm the integrality of human action and nonhuman agency in the making of local eco-climatic conditions. However, they seemingly collapse the important distinction between short-term weather events and long-term climate history, patterns, and change. Today, highlanders and theorists alike must account for not only structural changes in climate and environment but also the experienced failure, throughout the highlands, to combat these changes through ritual. Consequently, climate change—as opposed to weather change—is an existential conundrum that is making itself felt with increasing intensity in the highlands through the vanishing of glaciers and snow, the drying up of the soil, and more frequent natural disasters.

The problem of structural climate and environmental change, as opposed to temporal weather change, is the rising uncertainty about the procedures and effectiveness of ritual. Whereas, in the past, rituals occasionally failed to deliver the desired weather, welfare, and well-being, such failures were generally temporary in nature and could be reversed, within reasonable time, through different rites and other adjustments and interventions. By contrast, today's rituals do not fail occasionally or temporarily, but most of the time. This poses a problem of ontological dimensions. Increasingly often, highlanders confess to acting ritually without being certain whether the ritual will have the desired effect, or whether it is the appropriate action to take, despite its successful mediation of human–mountain mutuality over the course of many centuries. Ama Jomo, and other *yul lha*, never had a reputation for parsimony in response to ritual requests. Therefore, in addition to being worried about the younger generations' lack of enthusiasm for ritual, elderly highlanders are especially concerned that the rites they perform seem to be receiving a haphazard—or even no—response from Ama Jomo and other nonsecular beings.

In sum, in highland ways of being–relating–knowing, climate change is significantly also encountered through recurrent ritual failure that communicates a breakdown in mountain–human mutualism. It feeds a predicament of knowing–unknowing that is sustained by visiting scientists', development consultants', and government officers' obscure and far-fetched—to highlanders—explanations that global greenhouse gases, burning fossil fuels, and deforestation are causing Bhutan's mountains to change color. At the same time, the current era of socio-ecological unraveling and ritual uncertainty does not run completely contrary to highland knowledge traditions. Predictions and prophecies have long foretold the current discord and separation between mountains and humans. According to Sonam Dorji, the next, final stage is *zamling kabenob*—the end of the world—when all life on earth will burn under the heat of seven suns.

References

Allison, Elizabeth. 2019. "Deity Citadels: Sacred Sites of Bio-Cultural Resistance and Resilience in Bhutan." *Religions* 10(4): 1–17.

Aris, Michael. 1979. *Bhutan: The Early History of a Himalayan Kingdom*. Warminster: Aris & Phillips, Ltd.

Banerjee, Milinda and Jelle J.P. Wouters. 2022. *Subaltern Studies: 2.0. Being against the Capitalocene*. Chicago: Chicago University Press/Prickly Paradigm Press.

Choki, Kinley. 2021. "Cordyceps, Climate Change and Cosmological Imbalance in the Bhutan Highlands." In *Environmental Humanities in the New Himalayas: Symbiotic Indigeneity, Commoning, Sustainability*, edited by Dan Smyer Yü and Erik de Maaker, 152–166. Abingdon and New York: Routledge.

Cruikshank, Julie. 2005. *Do Glaciers Listen? Local Knowledge, Colonial Encounters, and Social Imagination*. Vancouver: University of British Columbia Press.

D'Avignon, Robyn. 2022. *A Ritual Geology: Gold and Subterranean Knowledge in Savanna West Africa*. Durham, NC: Duke University Press.

Dowdy, Sean. 2022. "The Bronze Sandal, or a Defense of Cosmic Refusal." *Political Theology*, July 27: 1–6. https://doi.org/10.1080/1462317X.2022.2095857.

de la Cadena, Marisol. 2015. *Earth Beings: Ecologies of Practice across Andean Worlds*. Durham, NC: Duke University Press.

Gagné, Karine. 2019. *Caring for Glaciers: Land, Animals, and Humanity in the Himalayas*. Seattle: University of Washington Press.

Gergan, Mabel D. 2017. "Living with Earthquakes and Angry Deities at the Himalayan Borderlands." *Annals of the American Association of Geographers* 107(2): 490–498.

Gohain, Swargajyoti. 2020. *Imagined Geographies in the Indo-Tibetan Borderlands: Culture, Politics, Place.* Amsterdam: Amsterdam University Press.

Haberman, David L. 2021. *Understanding Climate Change through Religious Lifeworlds.* Bloomington: Indiana University Press.

Huber, Toni. 1999. *The Cult of Pure Crystal Mountain: Popular Pilgrimage and Visionary Landscape in Southeast Tibet.* New York: Oxford University Press.

Huber, Toni. 2020. *Source of Life: Revitalisation Rites and Bon Shamans in Bhutan and the Eastern Himalayas.* Vienna: Austrian Academy of Sciences Press.

Huber, Toni and Poul Pedersen 1997. "Meteorological knowledge and Environmental Ideas in Traditional and Modern Societies: The Case of Tibet." *Journal of the Royal Anthropological Institute* 3(3): 577–597.

Ingold, Tim. 2000. *The Perception of the Environment: Essays in Livelihood, Dwelling and Skill.* London: Routledge.

Phuntsho, Karma. 2013. *The History of Bhutan.* New Delhi: Random House.

Phuntsho, Tashi. 1992. "Keep off the Mountain!" *Himal*, November 1. www.himalmag.com/keep-off-the-mountain/.

Pommaret, Francoise. 1994. "On Local and Mountain Deities in Bhutan." *Hal Open Science.* https://shs.hal.science/halshs-00717941/document.

Pommaret, Francoise. 2004. "*Yul* and *Yul Lha*: The Territory and its Deity in Bhutan." *Bulletin of Tibetology* 40(1): 39–67.

Rasmussen, Matthias Borg. 2015. *Andean Waterways: Resource Politics in Highland Peru.* Seattle: University of Washington Press.

Rasmussen, Matthias Borg. 2016. "Unsettling Times: Living with the Changing Horizons of the Peruvian Andes." *Latin American Perspectives* 43(4): 73–86. https://doi.org/10.1177/0094582X16637867.

Smyer Yü, Dan. 2020. "The Critical Zone as a Planetary Animist Sphere: Ethographing an Affective Consciousness of the Earth." *Journal for the Study of Religion, Nature and Culture* 14(2): 271–290. http://doi.org/10.1558/jsrnc.39680.

Smyer Yü, Dan. 2021. "Situating Environmental Humanities in the New Himalayas: An Introduction." In *Environmental Humanities in the New Himalayas: Symbiotic Indigeneity, Commoning, Sustainability*, edited by Dan Smyer Yü and Erik de Maaker, 1–21. Abingdon and New York: Routledge.

Smyer Yü, Dan and Jelle J.P. Wouters, eds. 2023. *Storying Multipolar Climes of the Himalaya, Andes, and Arctic: Anthropocenic Climate and Shapeshifting Watery Lifeworlds.* Abingdon and New York: Routledge.

Tashi, Kelzang T. 2023. *World of Worldly Gods: The Persistence and Transformation of Shamanic Bon in Buddhist Bhutan.* Oxford: Oxford University Press.

Ura, Karma. 2001. "Deities and Environment: A Four-Part Series." *Kuensel: Bhutan's National Newspaper*, November 26.

Winkler, Daniel. 2010. "*Cordysceps sinensis*: A Precious Parasitic Fungus Infecting Tibet." *Field Mycology* 11(2): 60–67.

Wouters, Jelle J.P. 2021. "Relatedness, trans-species knots and yak personhood in the Bhutan Highlands." In *Environmental Humanities in the New Himalayas: Symbiotic Indigeneity, Commoning, Sustainability*, edited by Dan Smyer Yü and Erik de Maaker, 27–42. Abingdon and New York: Routledge.

Wouters, Jelle J.P. 2023. "Where is the 'Geo'-Political? More-than-Human Politics, Polities, and Poetics in the Bhutan Highlands." In *Capital and Ecology Developmentalism, Subjectivity and the Alternative Life-Worlds*, edited by Rakhee Bhattacharya and G. Amarjit Sharma, 181–202. Abingdon and New York: Routledge.

11 Predatory Climes

Beastly Encounters in the Making of the Sundarbans[1]

Jason Cons

Jolil Mama is laughing at me. We are in Gabura, an island in Bangladesh's southwest delta zone on the fringe of both the Sundarbans and the India–Bangladesh border. We are sitting on Jolil's brother-in-law's terrace, drinking tea. Jolil works in the mangrove forests. An old *bawali*—literally wood collector, but broadly meaning forest worker—with a dashing air and somewhat unsavory reputation, he speaks with the authority of someone who has spent his life learning the lessons of the Sundarbans—often referred to simply as the "*jangal*"—the hard way. His livelihood has long depended on navigating the mutable boundaries between land and water, human and animal, legal and illegal, and predator and prey that characterize the Sundarbans—the world's largest remaining mangrove forest and one of the emergent ground zeros of global climate change.[2] He spends days, sometimes weeks, at a time under the mangrove canopy. It's a hard way to make a living. Jolil must navigate the vagaries of extracting resources from the mangrove forest (fish, crabs, honey, timber, smuggled goods from across the border with India). Changing weather patterns and salinity balance in the water make this harder, or at least less predictable. But Jolil also must contend with human and nonhuman predators under the mangrove canopy. These include forest officials enforcing policies that criminalize his livelihood in the name of protecting the Sundarbans; policing and paramilitary patrols seeking to make it safe for conservation; forest bandits who prowl its canals and waterways; and the Sundarbans tiger.

Jolil is one of my self-appointed tutors. Today, he has been schooling me on honey collecting. During the honey season, there's good money to be had hunting for giant hives in the mangrove depths, smoking out the fierce bees, and bringing back their honey to sell to brokers in villages along the forest fringe. Honey collecting is a dangerous occupation. Anyone who works in the jungle might fall prey to one of the dwindling number of tigers that stalk the mangroves, but honey collectors suffer a disproportionate number of attacks because their work demands that they enter the interiors of mangrove-dense islands—claustrophobic spaces where it can be hard to tell if you are being watched.

I ask Jolil about the techniques for avoiding tiger attacks. "What do you want to know?" he replies. I tell him that I have heard stories that in West Bengal, honey collectors used to wear a type of mask with a face on the back

DOI: 10.4324/9781003484394-11

of the head. When a tiger saw the face, they thought that they were being watched and do not attack. "Have you worn a mask while honey collecting?" I ask. Jolil stares at me for a few seconds, as though trying to gauge whether I am messing with him, and then bursts into uproarious laughter. He has seemingly never heard something so absurd. I expose myself as yet another foreigner visiting the Sundarbans with a head full of imaginations about both the tiger and the territory it stalks.

"Listen," Jolil says, "when you collect honey, you stay in groups, close together. Numbers are the only thing that will keep the tiger from attacking." Then, pulling up his lungi (the cloth skirt commonly worn by men in rural Bangladesh), he presents me with his leg. The bone is intact, but his calf is misshapen and smooth, covered in scar tissue. A chunk is missing. "This is what happens if you hunt alone." Jolil tells me about his encounter with the tiger that took a bite out of his calf. He and a group of companions were searching out honey, deep in the mangroves. He had forged ahead after hearing the buzzing of a large hive. That was when the tiger came out of the mangroves and seized his leg in its jaws. Hearing his cries, the rest of his team rushed forward and began striking the tiger with sticks and poles. It dropped Jolil and retreated into the forest to seek easier prey. Jolil drapes the lungi back over his leg and looks up at me. "Masks," he says, and begins laughing again.

Much has been written and said about the Sundarbans tiger, *Panthera tigris tigris*. It is so densely woven through discussions of the *jangal* that the two are all but synonymous. The tiger that stalks the Sundarbans is more than just a predator. It is also a deity, a symbol of sovereign power, an icon of national and international conservation, and a sentinel of global climate change. As my conversation with Jolil Mama suggests, my own imagination of the tiger leads me to misapprehend the nature of the beast. But the challenges that Jolil faces in his everyday struggles to make a living also suggest ways that imaginations like mine are profoundly entangled in the making of predation, ecology, and risk in the Sundarbans.

This volume considers the value of the concept of clime in thinking through the relationship between climate and place. It provides a useful alternative to frameworks for exploring global environmental change that see places as fundamentally subsumed by broader systems—where islands like Gabura become locales where climate change is experienced but not shaped.[3] Clime holds out a different possibility. It is not an emic term, at least in the contexts where I work. Although, as a concept that considers place in dialogue with its climate, it approaches the Bengali word *poribesh*—a complex term typically translated as "environment." Clime does highlight something of fundamental import to contemporary conversations about climate change. In the words of Dan Smyer Yü (2023, 8),

> when place is seen as clime, it is place imbued with climate agency, history, pattern, and change. It may appear to be a passive recipient of climate change understood as numerical global averages in climate science; however, in the sensible world, we recognize clime as an agent of climate change, too, an inherently living dynamic of the Earth.

To think of place from the standpoint of clime, then, is to begin to disaggregate the parts/whole relationship—to see every instantiation of time-place and every experience of socionature as both produced by and producing broader patterns of environmental and climate change. It is to put mutual constitution back on the table.

Clime—a region considered with reference to its climate—dissolves distinctions between instance and system, between phenomenology and science, and between local and global. It obviates the question of whether something is or is not climate change, and instead demands ethnographic attention to the more-than-human, more-than-sensory, and more-than-material constitution of place. It is a concept concerned with imbrication and entanglement—asking us to think past simple framings of, for example, fishermen in the Sundarbans as subject to an unknowable system. Here, everyday struggles to survive are bound up in the production of the Sundarbans' muddy clime—alongside and as much as the emissions of coal-fired power plants, the shifting monsoonal patterns that make up the South Asian hydro-cycle, and the negotiations of delegates at a Conference of the Parties. Here also, systems of knowing the Sundarbans—of navigating its currents, its tides, its sediments, its flora and fauna—intermingle with and occasionally cut across other scientific and governmental ways of knowing the delta's climate.

The problem of thinking through climes is one of vantage point. If we are to think past simple binaries of parts and whole, and abstract divisions of past, present, and future, where do we start? In this chapter, I suggest that thinking with tigers pushes the notion of clime in productive ways. It demands that we see tigers, and climes more broadly, as constituted through a murky admixture of environmental change, lived experience, and global imagination. Tigers have been a part of the delta landscape since at least the Late Pleistocene (over 12 millennia ago). They are more than just residents of the mangrove forests. As apex predators, they continue to play a critical role in balancing fauna, and consequently flora, in the region, allowing mangroves to flourish—to become the Sundarbans.[4] In this sense, tigers might be thought of as clime-makers.[5] But the Sundarbans tiger demands that we broaden our understanding of the concept of climes to show how such socionatural transformations are always produced through mediations that tie the happenings of place to processes elsewhere. As such, tigers are nodes through which one can see a multitude of articulations—between the local and global, the past and the future, the material and the symbolic. They highlight the ways that climes are the outcomes not simply of natural histories but also of capital flows and global discourses that invoke them to various ends. That is to say, they open a window on the predatory clime of the Sundarbans.

To tackle this challenge, I revisit a notable explanation of the local and the translocal Sundarbans: Annu Jalais's (2008) landmark essay "Unmasking the Cosmopolitan Tiger."[6] In this essay, Jalais, writing from the Indian side of the

Sundarbans, explores two conflicting versions of the tiger. The first—which she terms the "cosmopolitan tiger"—is the infinitely reproduced animal that appears in iconography, postage stamps, WWF logos, colonial fantasies, and so on. It is the tiger stripped of context and reduced to signifier—the image that can be appropriated for any meaning. It might stand for the ferocity of the animal kingdom, the nobility of the big cat, the sovereignty of the colonial or postcolonial state, the fragility of nature. It stands in contrast to what Jalais calls the "Sundarbans tiger"—the animal that fishermen like Jolil encounter, as it were, in the flesh.

Jalais's essay inquires into the troubled nature of the relationship between these two tigers. Her main concern is to suggest that the "cosmopolitan" vision of tigers, in her words, "restricts alternative spaces and ways of thinking about wild animals and their locales along non-Western-centric models" (Jalai 2008: 26). Moreover, Jalais suggests that the cosmopolitan tiger may be a threat to both its flesh-and-blood counterpart and to those who make human lives alongside them. As she points out, "universally propagated ideas about tigers ultimately act to the detriment of 'other' tigers because they do not allow an engagement with alternative ways of understanding animals and wildlife" (Jalai 2008, 26).

The collision between the cosmopolitan tiger and the Sundarbans tiger is a way for people who live in the Sundarbans to work out their own relationship with and mattering to the Indian state and international institutions concerned with tiger protection. That is to say, it highlights that the lives of peasants and fishermen who live in and on the fringe of the mangrove forests and work its swampy terrain matter significantly less than their beastly counterparts. But it also occludes understandings of kinship and multispecies mattering that offer potentially otherwise interpretations of environmental change.[7]

The implications of thinking the tiger as always both cosmopolitan and material help to open an important dimension in the discussion of climes. In climes, global imaginations of space are at play alongside humans, nonhumans, weather, land, water, and a changing climate. If a clime is a region understood with reference to its climate, then all its climates—environmental, intellectual, policy, development—matter. We need to think of these forces as bound together in ways that are not always easy to disentangle. Taken as such, beastly encounters in places like the Sundarbans are moments when we can begin to rethink the predatory politics of place and of the Anthropocene at large (Mathur 2021). In what follows, thinking from the Bangladesh side of the Sundarbans (the opposite side of the delta to Jalais) and from a moment of flux (the present), I explore the region's clime as, at least in part, mutually constituted with the flesh and figure of the tiger. I frame this as a predatory clime—a region characterized not solely by the predatory actions of a single entity, tigers, but by a broader ecology of capture and predation that figures tigers not only as predators but also as central objects of fascination and concern (see Cons 2021).

Muddy In-Between

It is hard to imagine a place more well suited to being understood as a clime than the Sundarbans. The mangroves are almost impossible to describe without constant reference to their climate. Nested within a broader climatic system that Amrith and Smyer Yü (2023) term the "Monsoonal Clime," it is a space composed of shifting terrain, shaped through a constant interplay of downstream flow from the Himalayas to the north and tides and storms from the Bay of Bengal to the south (Da Cunha 2019). These patterns produce the region as a zone of profound fluidity—where islands obstinately refuse to remain in place and rivers meander, where salt water (*labon pani*) seasonally switches to sweet water (*mishti pani*) and back. It is best understood as a space of modal indeterminacy where land, water, identity, religion, and more are constantly shifting. Here, binaries do not stand in stark opposition to one another, but rather seep and dissolve into each other across unstable boundaries (Cons 2020).

Scholars have written the modern history of the mangroves and the delta in which they sit as a story of human attempts to shore up the locations of land and water—to perform the Borgesian task of super-imposing the map onto the land.[8] This exercise in forging an ontological certainty about the relationship of wet to dry was necessary to the colonial and postcolonial project of imposing property relations to facilitate the extraction of revenue, taxes, and rents. Yet, this attempt to fix place in space constantly runs afoul of the hydrology of the delta. The result is neither landscape nor hydroscape but siltscape—an amphibious zone characterized by oscillations of weather, tide, and downstream flow.

As a clime, the Sundarbans is defined in relation to its broader environs. It is the obverse of the Himalayas—the low-lying, swampy counterpoint to the stark and jagged mountains to its north. At the same time, it is more than simply a space "downstream" from the mountains. The Bengal Delta and the mangroves at its mouth are a zone of mediation in the complex hydrological cycle that ties the mountain to floodplain to sea. On the one hand, it is the outcome of the millennia-long transfer of sediment from the Himalayas to the Bay of Bengal—the slow alluvial rendering of the alpine vertical as the delta horizontal. Bangladesh is, as Willem van Schendel (2020, 3) notes, "the Himalayas flattened out" (see also Amrith and Smyer Yü 2023). In this sense, the Sundarbans might be construed as the muddy feet of a greater Himalayan clime. On the other hand, it is a point of articulation in the broader monsoonal assemblage that returns moisture from the Bay of Bengal back to the mountains in the form of rain, glaciation, and snow (which subsequently carries more sediment from peak to bay).[9] As a delta, its silty terrain is defined through its connection to the upstream. But it also buffers inland space from the ocean's occasional fury. Its ecology reflects and refracts the regional and global dynamics of environmental change that are transforming ecologies upstream and down. On the one hand, the damming of upstream rivers (perhaps most notably India's Farakka Barrage) reduces the downstream flow of fresh water into the delta, making it a saltier and siltier space. On the other, the changing monsoonal patterns that accompany global warming expose the region to more frequent cyclones and storms even as ocean acidity changes the ecology of the forest, and what can and cannot thrive within it.[10]

If the Sundarbans is inextricably tied to other proximate and remote spaces, it is also fundamentally partitioned. The forest is bifurcated by the India–Bangladesh border—a line dating to Partition in 1947 that is a marker of both communal imaginations of territory and violence on the ground.[11] Today, the fence that India has constructed along the border is both a dramatic attempt to realize the demographic fantasy of creating Hindu and Muslim territory on either side and a site of extreme violence where those crossing frequently meet the lethal force of India's Border Security Force (BSF).[12] The fence does not reach into the swampy depths of the *jangal*, though this area is patrolled by the armed boats of both countries' security forces. If the border is a marker of the violence on the landscape, it also stands for the bureaucratic absurdity of managing nature across international boundaries. The division of this ecology into two nation-states has tremendous implications for its management. The region has two different national parks, two different delta plans, two different enforcement and policing strategies, and two different visions of what settlement in and around it in a future of global warming might mean.

Beyond the border, another division complicating this clime is the boundary of the national park itself. This also articulates differently on either side of the border. In India, the park sits within a broader Sundarbans region, where the ecology is not so simply divided into biosphere and settlement. In Bangladesh, the forest is hemmed in by densely settled islands that share its geography, ecology, and history. But they are largely denuded of mangroves and marked by settled agriculture, cleared land, and, since the beginning of a long boom in export-oriented shrimp aquaculture in the 1980s, endless expanses of shallow, saline shrimp *ghers* (ponds).[13]

This boundary also shapes global imaginations of the region and the tigers within it. Consider what is perhaps the most recognizable image of the Sundarbans—the Landsat 7 image of the Sundarbans reproduced in Figure 11.1. Landsat satellites are equipped with technology that renders the mangroves in dark green and the land in much lighter, brown–green shades to make it possible to track land-use change over time. Consequently, their images are often deployed in literature on global and regional climate change to invoke the fragility of nature and the urgent need for its protection.[14] Viewed from satellites without these spectral characteristics, the mangroves are much harder to distinguish from the land. When reproduced without explanation, the Landsat 7 image reinforces a classic conservation narrative of nature and culture: verdant nature besieged by development and degradation.

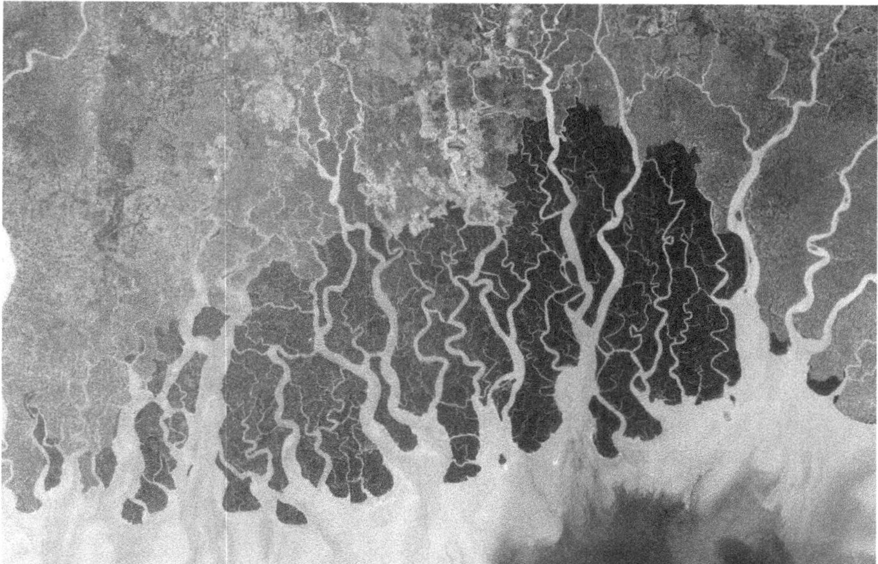

Figure 11.1 Landsat 7 image of the Sundarbans showing divisions between the forest and the land beyond
Note: NASA image created by Jesse Allen, Earth Observatory, using data obtained from the University of Maryland's Global Land Cover Facility.
Source: Wikimedia Commons: https://commons.wikimedia.org/wiki/File:Sundarbans.jpg?uselang=en#Licensing

Tigers are threaded through the delta, featuring prominently in its history, mythology, and endangered present. They occupy a central place in the cosmology of the Sundarbans—perhaps most notably as the physical manifestation of the deity Dakkhin Rai, the "King of the South," who is held at bay by Bonbibi, the syncretic mother of the forest who offers protection from tiger attacks and other misfortunes.[15] The colonial and postcolonial history of the tiger in the Sundarbans, and in South Asia at large, has been well documented by environmental historians. Much of this history might be construed as a productive tension between the Sundarbans tiger as noble—indeed, imperial—beast and tiger as man-eating threat. Indeed, this dialectic is at the heart of what Anand Pandian (2001) has called "predatory care"—the endlessly reproduced colonial fantasy of the paternal protection of local populations by white hunters ready to kill marauding man-eaters with European skill, bravery, and repeating rifles.[16] Man-eating tigers appeared as part of the colonial imagination of the Sundarbans, figuring prominently in framings of the mangroves as a sinister drowned land at the mouth of the delta (Greenough 2000).[17] In postcolonial South Asia, tigers have been reimagined as the public face of massive and well-known conservation efforts. This is most notable in India's Project Tiger, but tigers are equally central to Bangladeshi imaginations of conservation, threatened habitat, and imperatives to protect the mangrove forests for both national and global good (see Khan 2011).

Today, tigers in the Sundarbans are unquestionably under threat. The impact of climate change on them is a hotly debated topic, but most models note that rising sea levels and increased salinity (and decline of sweet water in the mangroves) are likely to erase viable tiger habitats in the near future.[18] Such changes in habitat, as explored below, are already shifting tiger behavior and patterns of predation—pushing them out of the mangroves and into settled communities on their fringe (Haque et al. 2015). Hunted by poachers for the lucrative trade in teeth and pelts, and facing the challenges of declining habitat and human predation, their survival has become both an open question and a preoccupation for those who seek to conserve the Sundarbans (Saif and MacMillan 2016). They thus also shape the postcolonial terrain in the delta, as the most prominent face of conservation and preservation of a mangrove forest, a global biosphere preserve, and a UNESCO World Heritage site (i.e., a place in need of preservation for the good of humankind at large).

Predatory Landscapes

I have never seen a tiger in the Sundarbans. The closest I came was a fresh pawprint, no more than a few minutes old, which I saw in 2020 (see Figure 11.2). Every time I complain about this to friends, they repeat a well-worn phrase,

Figure 11.2 The "just missed" tiger
Source: Photo by the author

"*tumi bagh dekhte pabe na, kintu bagh tomake dekhe*"—you won't see the tiger, but the tiger will see you. The phrase endows the tiger with panoptic power—a creature that instills fear and discipline on those who enter the *jangal*, demanding that you recognize the possibility that you might be subject to its lethal gaze at any moment. Still, most of my friends who work in the Sundarbans claim to have seen a tiger at least once. Many of the stories they share are vague and unremarkable, though some involve near misses and almost lethal encounters, like Jolil's. For all of that, most of the people I have encountered who work in the Sundarbans tend to see the tiger not as a malicious predator. Rather, they understand it as, on the one hand, a being that shares the space of the mangrove and, on the other, a form of risk beyond their immediate control.

Consider Shonkar, a crab fisherman and member of the Mondal community—an Adivasi group that straddles the India–Bangladesh border. One day, I ask him what it is like to fish for crabs in the mangroves. "In the nighttime, I live inside the *jangal*. In my boat, I am alone," he tells me. I comment that it seems like a solitary and dangerous way to make a living. "Honestly, if you are afraid, you cannot go into the Sundarbans," he says. "But, slowly, I have become used to it."

"What about tigers?" I ask.

"Of tigers, I have no fear. We Mandals have a mantra. I know this mantra by heart. If I say it, the tiger will not come in front of my eyes."

This attitude towards tigers—couched somewhere between devotion and the blasé—is one I've heard before. Some have mantras to mediate tiger risk. Others make offerings to Bonbibi. Piety offers protection to those who work in the forest. Or not. As a Muslim crab collector told me in response to a similar question, "If Allah wills it, my time has come."

Jalais explores a range of ways that those who live and work on the Indian side of the Sundarbans understand tiger attacks. Many of her interlocutors explain them as the result of transformations in the environment that have made it more violent. This violence has made the Sundarbans itself cantankerous—a space in which tiger, human, and tiger–human relations have consequently become more quarrelsome and potentially lethal. As Jalais (2008, 35) writes:

> Villagers explained that the growing violence of humans expressed through polluting paraphernalia such as motorboats, shrimpers' mosquito nets, and poachers' rifles, and more dangerous religious and political violence, affected the locale of the forest, which in turn affected tigers and other nonhumans' need for peace and security. This made tigers even more ferocious and increased the danger of working in the Sundarbans. The two (humans and nonhumans), however, are "sealed" together by this common environment of the Sundarbans—the locale of the Bengal tiger.

Jalais's narration is a remarkable explication of the region as clime. It situates human–tiger relations not as existing on opposite sides of an environmental binary (hunter/prey, human/nonhuman, nature/culture), but rather as forged within and by the same environ. Here, human and tiger life is thoroughly, if unevenly, intertwined with clime. In this view, tigers are, situationally, predators. But they are also only one predatory actor in the broader Sundarbans ecology. Others, both proximate and remote, are implicated in the remaking of this clime as one characterized by multiple forms of capture and violence. As the more-than-environmental climate of the Sundarbans shifts and becomes more violent, so does the more-than-human social climate of the delta. There is an intimate articulation between the violence *of* the environment and violence *in* it.

In their landmark volume *Violent Environments*, Peluso and Watts (2001) note that violent environments emerge not *a priori* out of scarcity, but rather out of social relations of production and social fields of power: political economies that can be understood only through a situated mapping of regimes of accumulation, modes of access to and control of resources, and the interplay of actors such as firms, peasants, and the state. Jalais's ethnography deepens this insight, expanding our understanding of the social relations of production by situating violent environments in broader patterns of change and centering the nonhuman as actor in their constitution. It demands at once that we take more-than-human (and more-than-biotic) agency seriously and that we center the experience and interpretations of those who make climes in our understanding of them.

I have asked friends on the other side of the delta about their interpretation of tiger encounters. Most offer prosaic accounts. They see tigers as unpredictable threats that cannot be planned for. Yet, they largely agree with Jalais's argument in that they understand of tiger attacks as articulating with and enmeshed in the increasingly violent environment. Tigers are afflicted by the same forces that shape the lives and livelihoods of humans who fish in the Sundarbans—increasing salinity in the water, conflict between armed bandits and the paramilitary forces that hunt them, pollutants that degrade the mangroves, the decline of available fauna for food, ever more hunting and poaching, unpredictable weather. While such transformations mean that my friends must spend more time under the mangrove canopy to capture sufficient resources to feed themselves and their families, it also means that tigers are leaving the canopy with increasing frequency to search for food in the densely populated agrarian space.

Since I began working in the delta region in 2015, there have been numerous reports of tigers emerging from the *jangal* to prey on domestic livestock. Tigers have long stalked the mangroves, moving from place to place, establishing their own predatory ranges, shifting locations with the seasons, weather, and storms. Yet, changes in ocean acidity, water levels, and salinity pose challenges both for accessing fresh water (also a challenge for humans living in the region) and for finding enough prey in their jungle habitats. Cows and goats owned by peasants living in villages near the Sundarbans have become regular supplements to tiger diets. Not surprisingly, there have been occasional accounts of villagers (often described as "mobs") attacking and killing these predators-out-of-place. Such accounts are often accompanied by lurid photos reproduced in local and national newspapers and YouTube videos displaying the tiger's dead body.

Jalais notes that the cosmopolitan nature of the tiger—its seemingly universal appeal and ubiquitous representation—produces a range of coercive relationships between wildlife in the forest and those who live in and along its borders. Human–tiger encounters in the Sundarbans are one such coercive relationship. The killing of tigers in villages on the fringe of the *jangal* is rendered as tragic in the international and national press—an avoidable catastrophe caused by peasants and fishermen who fail to understand the tiger's import (to everyone else). Such renderings radically simplify and erase other interpretations that highlight the more-than-human, and more-than-tiger, violence of the Sundarbans' shifting clime.

I'm sitting in the courtyard of a house in a village on the fringe of the Sundarbans. The decaying house speaks of earlier comparative wealth—the building is constructed from concrete and stone—but also of more recent financial decline. The family who live here have recently lost their cow to a tiger. Momin, the patriarch, explains what happened:

> The whole year long, I have been tying that cow and feeding her in our home. But after the recent rice harvest, I let her free to graze on the *dhan* [paddy straw]. After some time, I saw my cow had not returned home. We searched other houses but could not find the cow … Later, I got the news that the tiger had gotten her inside the jungle.

In many ways, Momin's narrative is a typical one. Over the past three decades, much of the delta's landscape—and arable land—has been converted into shallow, saltwater *ghers* due to a boom in *bagdachingri* (tiger prawn) aquaculture (Paprocki and Cons 2014). As a result, there has been a drastic decline in the amount of grazing land for livestock, so families like Momin's now often let their cows graze freely. While this enables cows to consume the leftovers of human harvests, it also tacitly allows them to enter the protected mangrove forest (unsupervised), where they can access a range of nutrient-rich flora, albeit illegally.

Cows, like the one lost by Momin's family, are often the most significant assets owned by peasants living on the fringe of the Sundarbans, so their loss can represent a catastrophic blow to family finances. In Momin's case, the loss was doubled, as the cow was pregnant and due to give birth within the next few days. As we talk, Rokeya, the aging matriarch of the family, repeatedly breaks into tears. "My heart is going to break," she tells me. "Today we are not eating anything." The government of Bangladesh runs an insurance scheme whereby people who lose livestock to tigers may receive modest compensation. However, eligibility rests on proving that the animal was not

Figure 11.3 Shrimp *ghers* in the delta's siltscape
Source: Photo by the author

grazing in the restricted space of the *jangal* at the time of the attack. That can be difficult. Even when families present evidence that their animals were indeed attacked outside the forest, suspicious authorities often assume otherwise. This family's loss was compounded by a fear of arrest for allowing their cattle to enter the mangroves illegally. "This morning, some people in the area threatened us," Rokeya tells me. "They say the government will make a case against us. Now, our hearts are in our mouths. Maybe they will capture us and take us to prison. What will happen, we cannot say."

Such anxieties and losses speak to the ways that the Sundarbans is a clime in which various vectors of control coalesce in ways that harm humans and nonhumans alike. As the political ecologies of production heralded by the shrimp boom degrade and erase agricultural and grazing land, the effects of pollution and global warming—the outcomes of their own interwoven social relations of production—erode forest habitats. As tiger prawns push livestock into the mangroves, other kinds of environmental degradation push tigers out. The effect is a blurring of the artificial boundary between forest and community that draws together humans living on the edge of the Sundarbans, animals within them, and those who police the boundary in new and corrosive configurations. The region's predatory clime emerges not only out of shifting patterns of weather but also out of a host of relations of predation and consumption—a warming climate that is forcing tigers out of the forest, Western markets that demand cheap seafood, conservation projects seeking to secure the future of the cosmopolitan tiger, local land politics, and more.

Sentinel Beasts

In a productive intervention in the debate over anticipatory governance, Keck and Lakoff (2013) suggest that analysts should turn their attention to sentinel devices. As they point out, "in the contemporary context of ecological anxiety, the sentinel has taken on an expanded meaning: it has come to describe living beings or technical devices that provide the first signs of an impending catastrophe." Given the preponderance of these sentinel devices in contemporary moments of ecological catastrophe (canaries in the climate coal mine), they ask "how the detection of threat can be made to have political force." As images of polar bears on shrinking icebergs—or emaciated tigers swimming among the mangroves—suggest, the answer to this question might depend, in part, on the charisma of the sentinel in question. Indeed, the promise of the sentinel is that it might produce the political force necessary to instigate a response (before it is too late). The cosmopolitan tiger has a long history of playing this role in the region (see Chakrabarti 2004). The prolific *figure* of the tiger, in marked contrast to the dwindling number of actual tigers, is present in almost every representation, project, and discussion of the Sundarbans, typically invoking the specter of imminent environmental collapse. In the contemporary moment, the cosmopolitan tiger has made a smooth transition from being the face of global conservation to one of the faces of global climate change.

It is thus not surprising that tigers are a topic of constant speculation in the region. Where they are (or are not) is a conversation that not only has an impact on village lives but also has implications for conservation budgets. Ironically, this fascination with the presence/absence of the Sundarbans tiger often serves to reinscribe the political division that splits the mangrove forest into two separate, state-controlled preserves. A constant question in national narratives is not just how many tigers are left but in which nation-state they reside. This manifests in Bangladesh not only in news reports and tiger census projects that constantly track the population of tigers within the nation-state (as opposed to entirety of the Sundarbans) but in rumors that imagine that tigers have themselves become bound up in state and boundary formation. There is constant anxiety in Bangladesh that there are more tigers on the Indian side of the border. This is often figured as the result of geopolitical intrigue, rather than any "natural" affinity for the Indian habitat. Several people have told me of a rumor that the Indian government has set up one-way gates in the Sundarbans. So, when border-crossing tigers walk through them, they cannot subsequently return to Bangladesh. Such rumors are far-fetched (why would tigers go through a gate in the unfenced mangroves in the first place?) and usually acknowledged as such by those who share them with me. But they do reflect the fraught relationship between the two states across the notoriously violent border fence that hems in most of Bangladesh.[19] They also speak to the import of the tiger not only to conservation in general but also to national pride and capital flows, particularly in the form of development.

NGOs working in the region—both international and local—have launched a number of tiger-oriented projects. Some of these, such as the dramatically titled "Wild Team," work hand in hand with the government of Bangladesh to implement tiger conservation and awareness schemes. These train local communities how to react if tigers stray out of the mangroves and into their villages. For example, "Village Tiger Response Teams" (VTRTs) are organized to scare off tigers, thus preventing lethal encounters in the event of human–tiger contact. Other NGOs build tigers into their portfolios in the form of popular programs such as "Tiger Widows Funds." Tiger widows have emerged as a somewhat peculiar object of fascination in international coverage of climate change, driven by countless stories of women who have lost their husbands to the charismatic predator. As such, programs assisting tiger widows signal to international funders that local NGOs (who often work as subcontractors for international organizations with their own development and tiger-oriented programs) are engaged with the famed forest denizen—or at least its victims.[20]

One day, I asked the director of a local NGO about a Tiger Widows Fund his organization had recently launched. He told me that it provided crucial financial support to the many women who had lost husbands to tigers in the region. Curious to meet some of the recipients, the following day, I journeyed to a remote village on the fringe of the Sundarbans, where I encountered six women who had received grants from the NGO to help develop alternative livelihoods in the absence of their husbands. Several of them had used the money to start local fish, shrimp, and crab businesses. Strikingly, most of the tiger attacks that had widowed these women occurred far in the past. The most recent attack had occurred seven years earlier, and some of them dated back more than a dozen years. Of course, such programs can provide crucial and life-altering support to families struck by tragedy. But the temporal distance between the husbands' deaths and the establishment of the fund for their widows highlights a paradox of life in predatory climes.

Tiger attacks are only one of many ways in which it is possible to die in the forest. For example, it is not uncommon to hear of women whose husbands have fallen victim to drowning or human violence perpetrated by armed bandits or paramilitary patrols (Cons 2021). Yet, while there is scrupulous accounting of tiger-attack deaths, it is almost impossible to gain accurate figures of deaths due to other causes in the mangroves. In the fascination with tiger widows, the mode of death—being killed by the forest's most famous predator—seems to matter more than the fact of death itself. Having their husbands killed by a jungle cat draws these women into a circuit that ties together tigers, communities living in the Sundarbans region, local development organizations, and global NGOs and donor organizations interested in sustainable livelihoods—especially as those livelihoods relate to climate change and conservation of the cosmopolitan beast. That is to say, it draws them into a broader and ongoing production of the region as a predatory clime.

Beastly Encounters

This chapter has argued that tigers offer a critical vantage point from which to unpack the making of climes. As I have attempted to show, cat and clime are intimately interwoven and can be understood neither as things imposed from afar nor as things that emerge through intimate encounter alone. Here, multiple visions of the tiger, climate, the material terrain of the mangrove forest and its surrounds, global visions of conservation, Western diets, capital flows, and more collide and combine in a murky admixture. Tiger widows become objects of development fascination by virtue of the form of their husbands' deaths; villagers become criminals for allowing their livestock to graze; tigers, in both their discursive and material forms, make and transgress space. Working from the standpoint of the tiger does not allow for an easy disentanglement of these forces. But it does point towards ways that diverse imaginations of the relationship between nature and culture are embroiled in coercive social relations on the ground.

Nayanika Mathur (2021), in her exploration of how big cats in India become "crooked," discusses the value of using beastly tales that center the kinds of encounters I have narrated in this chapter as a means of rethinking the Anthropocene. She writes, "Taken together, [beastly encounters] ground the Anthropocene within localized politics and ecosystems and can serve to relay the voices, imaginaries, and opinions of those people … who are already coping with the damaging consequences of climate change" (Mathur 2021, 12). Beastly tales can unsettle simplistic narratives that frame human–tiger encounters simply as conflict over declining resources and declining (animal) populations. They resituate such encounters as nexuses of a broad swath of predatory relations and multi-scalar politics.

These relations are as central to the making of climes as changing patterns of the monsoon or shifts in the downstream flow of rivers into a delta. Indeed, they cannot be easily disentangled from either. Both cosmopolitan tiger and the Sundarbans tiger—as well as the humans and nonhumans it encounters with occasionally lethal consequences—are part of a densely interwoven clime assemblage. That is to say, climes are an emergent category, always in the process of becoming. And in the Sundarbans, tigers, and the fishermen and livestock they encounter, both shape and are one of their emergent properties. Jalais's reading of the cosmopolitan tiger demands that we continue to inquire into the Western particularity of "wildlife"—that thing which must be protected, paradoxically, as the heritage of universal humankind. The need for such critique is no less urgent than when Jalais wrote her landmark essay in 2008. If her work shows us how the cosmopolitan beast threatened to overwrite otherwise interpretations of the relationships between humans and tigers, situating these imaginaries in the context of everyday practice can also show the ways in which the tiger-multiple continues to stand at the heart of predatory climes.

Notes

1 Parts of this chapter were published in different form in Cons (2023).
2 Indeed, the Sundarbans is a space of flux that puts the lie to the notion that such binaries are (or could be) discrete (see Cons 2020). On more than human predation in the Sundarbans, see my discussion of the mangrove forest as an ecology of capture (Cons 2021). By calling the Sundarbans, and the delta in which it sits, a "ground zero" of climate change, I mean to signal both the material transformations brought on by global warming and the discursive imaginings that facilitate anticipatory interventions in the delta to manage the present and future effects of climate change (see Cons 2018).
3 For one articulation of this, see Timothy Morton's (2013) theorization of "hyperobjects."
4 For an excellent overview of tigers in the Sundarbans, see Khan (2011).
5 In this sense, tigers may be thought of as similar to other clime-making, endangered apex species in South Asia, such as the one-horned rhinoceros in Kaziranga, Assam, or the Asiatic lion of the Gir Forest in Gujarat.
6 I focus on this essay instead of Jalais's equally excellent ethnography, *Forest of Tigers* (Jalais 2010), because it is here that Jalais forcefully mobilizes the concept of "cosmopolitanism" to examine the ways that global imaginaries produce a sort of alienation between the Sundarbans tigers and human residents.
7 On the notion of an "anthropology of the otherwise," see Povinelli (2022).
8 See Bhattacharyya 2018; Da Cunha 2019; and Lahiri-Dutt 2014.
9 See Bookhagen (2015). On the monsoon as assemblage, see Bremner (2022).
10 For a more robust account of the physical geography of the region, see Brammer (2014).
11 For an overview of this history, see van Schendel (2004).
12 On the violence of the India–Bangladesh fence, see Jones (2012).
13 For more on the influx of population into the region and the clearing of land to reclaim delta space, see Paprocki (2021), Iqbal (2010), and Lahiri-Dutt (2014).
14 For example, Figure 11.1 features on the cover of Amitav Ghosh's (2016) book, *The Great Derangement*.
15 Bonbibi, a constant presence in Bangladesh's delta region, is worshipped by Hindus and Muslims alike. It is common to see shrines to her in villages and in the courtyards of wealthier homes throughout the delta. For further discussion, see Jalais (2010) and Uddin (2019).
16 Vijaya Ramadas Mandala powerfully extends this critique, arguing that the colonial hunt—especially, though not exclusively, the tiger hunt—was central to colonial governance. As he notes, "What became established as mere recreational sport in the late eighteenth and early nineteenth centuries was later identified as critical to the continuation of colonial commercial and political functions and the extension of territorial control, particularly evident in the government's ruthless policy regarding forest populations and the utilitarian approach to wildlife conservation" (Mandala (2018, 1).
17 Mahesh Rangarajan (2012) notes that, in the early twentieth century, more than a third of all deaths due to tiger attacks in the entirety of British India occurred in the Sundarbans. He attributes this to two facts: within the mangroves, where the tiger is effectively a semi-aquatic creature, chance encounters were more common than in other South Asian tiger habitats; and the field of tiger predation was narrower.
18 See, for example, Mukul et al. (2019), who predict that tiger habitats will vanish by 2070.
19 Hossain et al. (2018) note that tiger densities are actually higher close to the India–Bangladesh border, possibly because of the frequency of border patrols that discourage poachers.
20 This is not surprising, given that organizations like USAID regularly elide the relationship between tiger conservation and sustainable development. For an example, see: www.youtube.com/watch?v=54FSzClYbUA.

References

Amrith, Sunil and Dan Smyer Yü. 2023. "The Himalaya and Monsoon Asia: Anthropocenic Climes since the 1800s." In *Storying Multipolar Climes of the Himalaya, Andes, and Arctic: Anthropocenic Climate and Shapeshifting Watery Lifeworlds*, edited by Dan Smyer Yü and Jelle J.P. Wouters, 29–51. Abingdon and New York: Routledge.

Bhattacharyya, Debjani. 2018. *Empire and Ecology in the Bengal Delta: The Making of Calcutta*. Cambridge: Cambridge University Press.

Bookhagen, Bodo. 2015. "Glaciers and Monsoon Systems." In *The Monsoons and Climate Change: Observing and Modeling*, edited by Leila Maria Véspoli de Carvalho and Charles Jones, 225–250. London: Springer.

Brammer, Hugh. 2014. "Bangladesh's Dynamic Coastal Regions and Sea-Level Rise." *Climate Risk Management* 1: 51–62.

Bremner, Lindsay, ed. 2022. *Monsoon as Method: Assembling Monsoonal Multiplicities*. London: Actar Press.

Chakrabarti, Ranjan. 2004. "Local People and the Global Tiger: An Environmental History of the Sundarbans." *Global Environment* 3: 72–95.

Cons, Jason. 2018. "Staging Climate Security: Resilience and Heterodystopia in the Bangladesh Borderlands." *Cultural Anthropology* 33(2): 266–294.

Cons, Jason. 2020. "Seepage: That which Oozes." In *Voluminous States: Sovereignty, Materiality, and the Territorial Imagination*, edited by Franck Billé, 204–216. Durham, NC: Duke University Press.

Cons, Jason. 2021. "Ecologies of Capture in Bangladesh's Sundarbans: Predations on a Climate Frontier." *American Ethnologist* 48(3): 245–259.

Cons, Jason. 2023. "Delta Temporalities: Choked and Tangled Futures in the Sundarbans." *Ethnos* 88: 308–329.

Da Cunha, Dilip. 2019. *The Invention of Rivers: Alexander's Eye and Ganga's Descent*. Philadelphia: University of Pennsylvania Press.

Ghosh, Amitav. 2016. *The Great Derangement: Climate Change and the Unthinkable*. Chicago: University of Chicago Press.

Greenough, Paul. 2000. "Hunter's Drowned Land: A Science Fantasy of the Bengal Sundarbans." In *Nature and the Orient*, edited by Richard Grove, Vinita Damodoran, and Satpal Sangwan, 237–272. New Delhi: Oxford University Press.

Haque, Mohammad Zahirul, Mohammad Reza, Sahibin Abd Rahim, Pauzi Abdullah, Rahmah Elfithri, and Mazlin Mokhtar. 2015. "Behavioral Change Due to Climate Change Effects Accelerate Tiger Human Conflict: A Study on Sundarbans Mangrove Forest, Bangladesh." *International Journal of Conservation Science* 6(4): 669–684.

Hossain, A.N.M., A.J. Lynam, D. Ngoprasert, A. Barlow, C.G. Barlow, and T. Savini. 2018. "Identifying Landscape Factors Affecting Tiger Decline in the Bangladesh Sundarbans." *Global Ecology and Conservation* 13: 1–10.

Independent BD. 2019. "Tigers May Go Extinct in 50 Years: Climate Change in the Sundarbans." February 15.

Iqbal, Iftekhar. 2010. *The Bengal Delta: Ecology, State and Social Change, 1840–1943*. London: Palgrave Macmillan.

Jalais, Annu. 2008. "Unmasking the Cosmopolitan Tiger." *Nature and Culture* 3(1): 25–40.

Jalais, Annu. 2010. *Forest of Tigers: People, Politics, and Environment in the Sundarbans*. New Delhi: Routledge.

Jones, Reece. 2012. *Border Walls: Security and the War on Terror in the United States, India, and Israel*. London: Zed Books.

Keck, Frédéric and Andrew Lakoff. 2013. "Preface: Sentinel Devices." *Limn* 3. https://limn.it/articles/preface-sentinel-devices-2/.

Khan, M. Monirul. 2011. *Tigers in the Mangroves: Research and Conservation of the Tiger in the Sundarbans of Bangladesh*. Dhaka: Arannayk Foundation.

Lahiri-Dutt, Kuntala. 2014. "Beyond the Water–Land Binary in Geography: Water/Lands of Bengal Re-Visioning Hybridity." *Acme* 13(3): 505–529.

Mandala, Vijay Ramadas. 2018. *Shooting a Tiger: Big-Game Hunting and Conservation in Colonial India*. New Delhi: Oxford University Press.

Mathur, Nayanika. 2021. *Crooked Cats: Beastly Encounters in the Anthropocene*. Chicago: University of Chicago Press.

Morton, Timothy. 2013. *Hyperobjects: Philosophy and Ecology at the End of the World*. Minneapolis: University of Minnesota Press.

Mukul, Sharif A. et al. 2019. "Combined Effects of Climate Change and Sea-Level Rise Project Dramatic Habitat Loss of the Globally Endangered Bengal Tiger in the Bangladesh Sundarbans." *Science of the Total Environment* 663: 830–840.

Pandian, Anand. 2001. "Predatory Care: The Imperial Hunt in Mughal and British India." *Journal of Historical Sociology* 14(1): 79–107.

Paprocki, Kasia. 2021. *Threatening Dystopias: The Global Politics of Climate Change Adaptation in Bangladesh*. Ithaca: Cornell University Press.

Paprocki, Kasia and Jason Cons. 2014. "Life in a Shrimp-Zone: Aqua- and Other Cultures in Bangladesh's Coastal Landscape." *Journal of Peasant Studies* 41(6): 1109–1130.

Peluso, Nancy and Michael Watts. 2001. "Violent Environments." In *Violent Environments*, edited by Nancy Peluso and Michael Watts, 3–38. Ithaca: Cornell University Press.

Povinelli, Elizabeth. 2022. *Routes/Worlds*. London: Sternberg Press.

Rangarajan, Mahesh. 2012. "The Raj and the Natural World: The Campaign against 'Dangerous Beasts' in Colonial India, 1875–1925." In *India's Environmental History*, Volume 2: *Colonialism, Modernity, and the Nation*, edited by Mahesh Rangarajan and K. Sivaramakrishnan, 95–142. New Delhi: Permanent Black Press.

Saif, Samia and Douglas MacMillan. 2016. "Poaching, Trade, and Consumption of Tiger Parts in the Bangladesh Sundarbans." In *The Geography of Environmental Crime*, edited by Gary R. Potter, Angus Nurse, and Matthew Hall, 13–32. London: Palgrave Macmillan.

Smyer Yü, Dan. 2023. "Multipolar Clime Studies of the Anthropocenic Himalaya, Andes and Arctic: An Introduction." In *Storying Multipolar Climes of the Himalaya, Andes, and Arctic: Anthropogenic Climate and Shapeshifting Watery Worlds*, edited by Dan Smyer Yü and Jelle J.P. Wouters, 1–26. Abingdon and New York: Routledge.

Uddin, Sufia. 2019. "Religion, Nature, and Life in the Sundarbans." *Asian Ethnography* 78(2): 289–309.

van Schendel, Willem. 2004. *The Bengal Borderlands: Beyond State and Nation in South Asia*. London: Anthem Press.

van Schendel, Willem. 2020. *A History of Bangladesh*. Cambridge: Cambridge University Press.

12 Afterword

A Himalayan–Andean Conversation

Karsten Paerregaard

Reading this volume's eleven chapters provides an eye-opening glimpse of a region that is historically, geopolitically, and culturally pivotal in an Asian context, but which in other parts of the world is associated with the spectacular heights of its mountains and the exotic customs and religious practices of its people. For an Andean scholar familiar with the challenges mountains pose to humans, but with little ethnographic knowledge of Himalayan society and culture, this volume is a Pandora's box of both expected and unforeseen surprises. Constituting the planet's two highest mountain ranges, the Himalayas and the Andes share multiple commonalities, not only with regard to physical environments, weather conditions, topographies, landscapes, and water bodies, but also with regard to that which concerns human adaptations, societal organization, livelihoods, and cultural ideas and practices. Living on the roof of the world (as the Himalayas are popularly known) or the ceiling of the world (as the Andes are accordingly labeled) demands unique skills to make use of the resources available while preserving human/nonhuman symbiosis. Indeed, one of the features that make the lifeworlds of the two mountain regions comparable is the intricate relation of reciprocity that embeds society in nature and that both Himalayan and Andean peoples ritualize in local folklore, cultural performativity, and spiritual imagination (Bolin 1998; Isbell 1978).

As an anthropology freshman, I was drawn to these wonders of mountain cultures that incited me to read all the books about Tibet, Nepal, Sikkim, and Bhutan I could find. My Himalayan curiosity was later fueled by movies, documentaries, and anthropological accounts of the region, and a few years ago I seized the opportunity to attend a workshop on climate change in Kathmandu. Over and over, I have been struck by what appear to be Himalayan and Andean similarities: sparkling glaciers and snowcaps covering breathtaking summits and reaching out for the blue sky; picturesque, turquoise mountain lakes decorating the barren and bleak landscape above the treeline; large water basins and raging rivers making their way from the mountain peaks to the ocean, where they create deltas with thriving habitats for plants, animals, and humans; and, what matters most to the anthropologist, villages on steep slopes and in deep valleys surrounded by human-made terraces with people dressed in colorful garments working the fields and

DOI: 10.4324/9781003484394-12

conducting vibrant ceremonies to honor the animated environment. Yet, over time, I have also learned that the more you dig into the details of what appear to be similarities, the more the regions' differences come to the fore, particularly regarding their political and social histories. Comprising an impressive selection of themes and aspects of the Himalayan lifeworld, the chapters of the present volume offer a singular invitation to compare the two regions—an exercise that global climate change makes especially timely. As Jelle J.P. Wouters points out in Chapter 1, the vulnerability of high mountains is often overlooked when attention is drawn to the impact of climate change in Arctic environments and the melting of the world's glaciers and icecaps (Orlove, Wiegandt, and Luckman 2008).

The Himalayas and the Andes are both folded or overthrust mountains formed by the compression of two tectonic plates. Moreover, both regions are chains of large mountains. As one of the planet's youngest mountain ranges, the Himalayas host several of its highest summits, including Mount Everest and K2. The Andes, by contrast, are not only much older but also the world's longest mountain range, stretching from Venezuela to Chile. Another important distinction is that the Andes are full of volcanoes, while the Himalayas have none—a difference due to the way the tectonic plates interact in the two regions. Geographically, the two mountain ranges lie in the tropics and subtropics, but because of their heights, they both comprise all four weather zones and are rich in biodiversity, with dry, cold climates on their summits that produce ice, snow, and lots of meltwater, and hot, fertile habitats at their feet that are fed by some of the world's largest and longest rivers. The nearby oceans and the wind and rain they produce have a crucial impact on the climate and environment of each region, albeit with different outcomes.

In the eastern Himalayas, the weather—especially precipitation—is synonymous with the monsoon, a topic Rima Kalita addresses in Chapter 6. Kalita shows how this southeasterly wind, which blows from the Indian Ocean and divides the Himalayan year into four seasons, has historically shaped human as well as nonhuman life in the Indian state of Assam. Unlike the western Himalayas, which are arid, the eastern Himalayas are green and fertile thanks to the monsoon, although people never know whether to expect deluge or drought, which has given the weather system its nickname: "the moody monsoon." Less blessed by rain than the Himalayas, but equally exposed to the weather's "moods," the Andean year has only two seasons—wet and dry—due to the Humboldt current, which flows along the eastern Pacific coast from the continent's southern tip to northern Peru. The cold water is so rich in fish that Peru is the world's second-largest fishing nation. However, the current produces little precipitation in the western Andes, so the region is dry and barren in comparison with the eastern Andes, which receive more rain and are consequently greener. Moreover, unlike the monsoon, whose windy power in the eastern Himalayas is undisputed (is that why it is "moody"?), the Humboldt periodically yields to a warm-current trend called El Niño that flows from the western to the eastern Pacific and every five to seven years turns the region's weather upside down.

Rain is associated with divine power throughout the world, including in mountain regions, where weather conditions are often unstable and where rain sometimes falls in excess (as occasionally happens in the Himalayas) and at other times arrives late or fails altogether (as often happens in the Andes). In Chapter 7, Sangay Tamang relates that the inhabitants of the Darjeeling Hills, a Himalayan foreland, ascribe to rain an agential power of the earth with the capacity to both generate and destroy life—a sacred quality that is shared by bodies of water, such as lakes, springs, and rivers. Though less abundant than in the Darjeeling Hills, rain is also essential for agriculture in the eastern Andes. Conversely, in the western Andes, meltwater from the mountains' glaciers and snowcaps constitutes most communities' principal water supply. Consequently, many villages pay tribute to their water sources and irrigation canals in annual ceremonies (Gelles 2000). Indeed, Andean people never take precipitation for granted, and in El Niño years, when it rains excessively in the north but very little in the south, they prepare for the worst. This is because they understand the power of water, which they regard as a living substance that gives and takes, sometimes in the same blow. In short, whether manifested in the form of rain, ice, or snow, water embodies the essence of both life and death in both the Himalayas and the Andes (Stensrud 2021).

The climatic variety and water resources of the Himalayas and the Andes provide a unique environment for animals and plants, some of which are indigenous and found only in the two mountain regions. Furthermore, as Roderick Wijunamai explains in Chapter 2, even though some plants, such as rice, originated in China and are widespread throughout Asia, in India's Southern Naga Hills, which lie at the foot of the Himalayas, *meusan* (freshly harvested rice) is considered native and inextricably connected with the local population's cultural identity. In fact, as Wijunamai points out, *meusan* is perceived as a gift. Unlike the Himalayas, which played host to many ancient civilizations and premodern empires, the Andes lie on a continent that was isolated from the rest of the world for thousands of years. Its remoteness explains why it is home to several crops that were unknown to the Spanish conquerors when they arrived more than 500 years ago (Crosby 1972). Today, the likes of maize, potatoes, avocado, and quinoa have become basic ingredients in the diets of billions of people around the globe. Yet, some of these crops are still considered the quintessence of Andean life. For instance, in the Andes, maize plays a critical role not only in people's daily nutrition but also in their offerings to deities. Likewise, according to legend, the favorite dish of the region's first inhabitants was quinoa porridge. As the saying goes, you are what you eat, and consuming this native grain reminds today's generations of their ancestors. Finally, the leaves of coca—another native Andean plant—are used to aid sociality, diagnose disease, and communicate with nonhuman agents (Allen 1988).

Animals also constitute a distinctive feature of the two regions. Again, the Andes stand out as an isolated setting for a host of native species, such as the condor and the guinea pig, as well as llamas, alpacas, vicuñas, and guanacos.

The Himalayas also have many indigenous animals, including the snow leopard, the yak, and the panda. Moreover, some predators have a particular impact on the region even though they are also found elsewhere in Asia. One of these is the Bengal tiger, which Jason Cons (Chapter 11) describes as a creature of majestic power. Living in the Sundarbans, the world's largest mangrove forest, where the Ganges and Brahmaputra merge, the tiger is both feared and respected by the other inhabitants. By contrast, the predators of the Andes do not represent much of a threat to humans, and although the puma and the bear are important figures in Andean cultural history, they do not have the same iconic status as the tiger. One of the region's most prominent animals is the condor (Bastien 1978)—a species of vulture that scavenges on carrion and thus cannot technically be classified as a predator. Even so, when the condor stretches its huge wings, soars on the air currents, and glides through the deep canyons of the Andes in search of food, it reigns as a sovereign, much like the tiger in the Sundarbans or the Himalayan snow leopard. Its royal status is underscored by its ritual role as a symbol of Andean resistance against the Spanish conquerors and a native counterpoint to Iberian culture (Murra, Watchel, and Revel 1986).

Moving from the nonhuman to the human world, comparing the Himalayas with the Andes is a showcase for those who claim that culture is the means by which humans adapt to and construe the physical world they inhabit. In the cryosphere of the Himalayas' Critical Zone, the herders of Bhutan's Thimphu highlands dwell in a world of water bodies that feed an archipelago of lakes and rivers which Jelle J.P. Wouters and Thinley Dema (Chapter 3) describe as "more-than-water" and a "terrestrial ocean." Viewing Bhutan as an aquatic land, they offer an account of *tshomen*, which the herders envision as mistresses of mountain lakes with the power to control the water flow. Engaging with these mermaids is pivotal to the herders' (as well as other Bhutanese's) well-being, much like Andean people's interactions with mountain deities known as *apu*, who are perceived as critical for humans' health and prosperity (de la Cadena 2015). And just as Bhutanese herders live with the danger of *tshomen* anger, Andean people sometimes fall victim to the lake mermaids' retribution. These are found all over the Andes, but one lake stands out not only because of its size (the largest by surface area in South America) but also because of its legendary status. According to Andean mythology, Lake Titicaca was the incubator of the first Incas (Urton 1981). However, as in Bhutan, more recent development and modernity are upsetting the mutual respect between humans and the nonhuman agents of Andean water bodies, drawing attention to the anthropogenic future that jeopardizes the region (Andolina, Laurie, and Radcliffe 2009; Boelens 2015).

Extending the focus on Bhutan's nonhuman world from lakes to the entire landscape, in Chapter 4 Kinley Dorji introduces us to storied toponyms and how they frame relations between humans and nonhumans, shape people's spiritual encounters and cultural memories, and produce place as not only a physical but also an affective locality. By using toponymic narration as a means of inscribing human–nonhuman relations into Bhutan's ecology, and approaching this as an ethical as well as a sentient universe, Dorji enables the reader to understand how people living in mountain regions account for climate change as more than an alteration of the earth's physical systems. Scientists convincingly argue that climate change is a global phenomenon. Yet, most people experience it as something that happens locally and read it into their own lifeworld. People living in mountain regions are no exception and, similar to the residents of Bhutan, Andean communities observe and interpret the environmental changes they are experiencing as a transformation of the places and localities they know and which they have mapped in toponymic stories. Accordingly, the names of mountains, rocks, lakes, and rivers embody their collective histories, as is evident when they walk along the borders of their communities to affirm their territorial rights each year. However, in both the Himalayas and the Andes, the toponymic stories need to be retold if name and places no longer match. Andean people have previously experienced historical changes with catastrophic consequences for their lives, which they later reformulated and incorporated in new narratives (Abercrombie 1998; Seligmann and Fine-Dare 2019). Can they do so again with regard to climate change? Can Bhutanese people do the same?

In Chapter 10, on the entanglement of herders and mountains in the Bhutan highlands, Deki Yangzom and Jelle J.P. Wouters explore how the Bon–Buddhist tradition frames Bhutanese genealogies of knowledge and praxis and discuss how these epistemologies and experiences both challenge and complement dominant scientific taxonomies of life and nonlife. Similarly, they show how climate change affects images of mountains and their deities, and how the nonhuman agency of Bhutan's landscape intersects with and molds climate change. Moreover, they point out that failure to communicate with nonhuman agents through rituals leads to a breakdown in mountain–human mutualism and unravels the way climate change is perceived and explained. Climate change threatens society and culture in the Andes, too—perhaps even more so than in Bhutan (Carey 2010). The Peruvian highlands house more than 70 percent of the planet's tropical glaciers, but these are melting at an alarming speed, causing not only serious environmental problems but also an ontological crisis as the *apu* and other nonsecular agentive life-forces either fail to respond to the communities' ritual offerings or, even worse, turn angry and make humans suffer from their misdeeds. From the Andes we also learn that, paradoxically, people who live in the poorest parts of the world and who are most vulnerable to global climate change often blame themselves rather than others for the degradation of their environment (Paerregaard 2023). We see the same thing happening in Bhutan (Chapters 3 and 10, this volume). Interestingly, in both regions, the mountains' lack of response to efforts to appease them has prompted people to intensify their ritual practices, rather than abandon them, in the hope that this will change their fortunes. This may be a sign that both Andean and Bhutanese people are incorporating climate change within their cosmologies by assuming responsibility for the damage it is causing to their lives and livelihoods.

A pilgrimage involves walking through unfamiliar environments, sometimes for days or weeks, enduring considerable suffering and pain, in order to reach a remote place regarded as sacred. Mountains offer the perfect setting for such an endeavor. In Chapter 8, Ainslie Murray recounts how pilgrims ascend the rivers of the Garhwal Himalayas to reach the sources of the Ganges, portraying their journeys as a motion in mountains as well as mountains in motion. Obviously, the pilgrims struggle to make their way up and down the Deep Time structures of the Himalayas, but the mountains also feel the impact of the pilgrims, especially as these increase in number, generating a massive collective body of people, each of whom leaves a personal imprint on the landscape. Apart from the different belief systems that drive them (primarily Hindu/Buddhist versus Andean/Christian), pilgrimages in the Himalayas and the Andes, and the physical effort and spiritual experience they entail, vary little (Sallnow 1987). Throughout the year, Andean pilgrims walk in their thousands to the summits of the region's mountains; and as the glaciers and icescapes retreat due to climate change, ever more people, predominantly driven by a New Age mindset, follow them to witness the environmental drama. Ironically, the impact of their activities is exacerbating the pressure on the mountains' fragile ecologies, which has compelled regional and national authorities in Peru to monitor and regulate the number of visitors. As Murray reminds us, pilgrims move in mountains but mountains also move in response. One may fear that, as they move the mountains, pilgrims forget what drove them to move themselves in the first place.

The aim of Himalayan pilgrims is to honor the sacredness of the region's mountains and, importantly, its water sources, including glaciers, snowcaps, springs, and rivers. The goal of mountain tourists, on the other hand, is to climb and "conquer" summits, to demonstrate humans' superiority over mountains—preferably the highest of them all, Nepal's Mount Everest. In Chapter 5, Jolynna Sinanan takes us on a climbing adventure to the mountain's cryosphere, as well as on a communicative journey through which the climbers and their guides disseminate their adventures to a global network of admirers. Sinanan's point is that modern communication technologies, and the digital practices and visual cultures they create, not only transmit an image of Everest ice- and mountain-scapes—or a "cryo-visual," as the author labels it—but also lay bare the commodification of the mountain's ecosystems and reveal the social and political economy of the tourist industry and the climbers' use of local labor. Likewise, in the past two decades, the Andes have become a major tourist destination in South America. Famous for attractions such as Machu Picchu, Peru received more than 5 million tourists annually before the COVID-19 pandemic. While most of the visitors are backpackers and sightseeing tourists, a significant proportion engage in mountain trekking and climbing, although only a few reach the cryosphere. Nevertheless, the anthropogenic effect of Peru's tourist industry is comparable to that of Nepal. In many Andean towns and communities, tourism is accelerating an existing water crisis, and forcing thousands of families into dependence on an industry that may crumble at any moment—as occurred during the pandemic. Truly, mountains are no longer a godly world out of human reach; rather, they have become a rallying point where pilgrims, tourists, and an entire digital universe celebrate the Anthropocene.

Whether through their geology, geography, hydrology, biology, zoology, mythology, folklore, ontology, religion, or tourism, the Himalayas and the Andes share many commonalities. Yet, when it comes to their sociopolitical histories, they differ in several important respects. In Chapter 9, Alexander Davis discusses the geopolitics of the eastern Himalayas, focusing on the current border conflict between India and China and its nexus with the Brahmaputra's riverine cosmos. Davis contextualizes the complexity of this dispute over the damming of the Brahmaputra and its tributaries by reviewing previous attempts in both imperial and postcolonial times to use the river to draw borders, and by examining the many conflicting socioeconomic interests that have driven such endeavors. Untangling the great challenges posed by attempts to engineer the river, Davis asks whether the two countries' engineers, militaries, and politicians will ever achieve their goals or whether the region's Deep Time history and the "geological surprises" it conceals will wash away their geopolitics. The Andes is also full of geological surprises. The region is rich in minerals and other natural resources, and extractivism has been the motor of economic growth in many Andean countries for centuries (Ødegaard and Rivera Andía 2019). Mining, in particular, has been a source of the region's recurrent social and political tensions, which often have roots in the continent's colonial history (Li 2015). Latin America gained independence two centuries ago, but the continent still struggles to overcome the demographic, political, and cultural changes that the conquest and subsequent Iberian rule entailed. Consequently, contemporary Latin American society is hierarchical—economically, socially, and ethnically (Gotkowitz 2011; van den Berghe and Primov 1977)—physically mobile, with high rates of rural–urban and international migration (Ødegaard 2020; Paerregaard 1997; 2015), and hybridized, with a cultural mixture of the continent's precolonial past and its long immigration history (Paerregaard 2008). Thus, even though Andean culture today displays multiple Inca and pre-Inca traits and attributes, these are inseparable from the syncretism that emerged from the indigenous population's encounter with Spanish conquerors, enslaved Africans, and Chinese and Japanese indentured workers (Jacobsen and Aljovín de Losada 2005).

As this volume shows, a clime approach contributes to a better understanding of the Himalayas' cultural and multispecies history. The same could be said for the Andes. As mentioned, unlike the Himalayas, which have been exposed to foreign contact and intervention for thousands of years, the region was isolated from the rest of the world until the Spaniards arrived, which is why it has so many native animals and plants, and why it is home to so many indigenous cultures. The conquest brutally ended this isolation and heralded a new era of death, loss, and uprooting, and later of cultural hybridization and racial mixture (Cook 1981; Stern 1983; Ramirez 1996). By contrast, Nepal, Bhutan, and (until recently) Sikkim and Tibet were never colonized, and while the outside world regularly left its signature on Himalayan environments, cultures, and religions, its people view these as aspects of their own, autonomous, unconquered lifeworld.

But history has no ending, and as globalization, tourism, geopolitics, and, more urgently, global climate change transform and impact the world's roof and ceiling, Himalayan and Andean people may soon find themselves facing similar challenges: rural–urban and international migration, digital communication, a new world order, global warming, and melting glaciers among them. There are, then, many more topics and questions to address in another volume on mountain societies and culture.

References

Abercrombie, Thomas. 1998. *Pathways of Memory and Power. Ethnography and History among an Andean People.* Madison: University of Wisconsin Press.

Allen, Catherine, J. 1988. *The Hold Life Has: Coca and Cultural Identity in an Andean Community.* Washington, DC: Smithsonian Institution Press.

Andolina, Robert, Nina Laurie, and Sarah Radcliffe. 2009. *Indigenous Development in the Andes: Culture, Power, and Transnationalism.* Durham, NC: Duke University Press.

Bastien, Joseph. 1978. *Mountain of the Condor: Metaphor and Ritual in an Andean Ayllu.* Prospect Heights: Waveland.

Boelens, Rudgerd. 2015. *Water, Power, and Identity: The Cultural Politics of Water in the Andes.* Abingdon and New York: Routledge.

Bolin, Inge. 1998. *Rituals of Respect: The Secret of Survival in the High Peruvian Andes.* Austin: University of Texas Press.

Carey, Mark. 2010. *In the Shadow of Melting Glaciers: Climate Change and Andean Society.* Oxford: Oxford University Press.

Cook, David Noble. 1981. *Demographic Collapse: Indian Peru, 1520–1620.* Cambridge: Cambridge University Press.

Crosby, Alfred W. 1972. *The Columbian Exchange: Biological and Cultural Consequences of 1492.* Westport, CT: Greenwood Press.

de la Cadena, Marisol. 2015. *Earth Beings: Ecologies of Practice across Andean Worlds.* Durham, NC: Duke University Press.

Gelles, Paul H. 2000. *Water and Power in Highland Peru: The Cultural Politics of Irrigation and Development.* New Brunswick: Rutgers University Press.

Gotkowitz, Laura, ed. 2011. *Histories of Race and Racism: The Andes and Mesoamerica from Colonial Times to the Present.* Durham, NC: Duke University Press.

Isbell, Billie Jean. 1978. *To Defend Ourselves: Ecology and Ritual in an Andean Village.* Prospect Heights: Waveland.

Jacobsen, Nils and Cristóbal Aljovín de Losada. 2005. *Political Cultures in the Andes, 1750–1950.* Durham, NC: Duke University Press.

Kroegel, Alison. 2011. *Food, Power, and Resistance in the Andes: Exploring Quechua Verbal and Visual Narratives.* Lanham: Lexington Books.

Li, Fabiana. 2015. *Unearthing Conflict: Corporate Mining, Activism, and Expertise in Peru.* Durham, NC: Duke University Press.

Murra, John, Nathan Watchel, and Jacques Revel, eds. 1986. *Anthropological Histories of Andean Polities.* Cambridge: Cambridge University Press.

Ødegaard, Cecilie Vindal. 2020. *Mobility, Markets, and Indigenous Socialities: Contemporary Migration in the Peruvian Andes.* Abingdon and New York: Routledge.

Ødegaard, Cecilie Vindal and Juan Rivera Andía, eds. 2019. *Indigenous Life Projects and Extractivism: Ethnographies from South America.* Cham: Palgrave Macmillan.

Orlove, Ben, Ellen Wiegandt, and Brian H. Luckman, eds. 2008. *Darkening Peaks: Glacier Retreat, Science, and Society.* Berkeley: University of California Press.

Paerregaard, Karsten. 1997. *Linking Separate Worlds: Urban Migrants and Rural Lives in Peru.* Oxford: Berg.

Paerregaard, Karsten. 2008. *Peruvians Dispersed: A Global Ethnography of Migration.* Lanham: Lexington Books.

Paerregaard, Karsten. 2015. *Return to Sender: The Moral Economy of Remittances in Peruvian Migration.* Berkeley: University of California Press.

Paerregaard, Karsten. 2023. *Andean Meltdown: A Climate Ethnography of Water, Power, and Culture in Peru.* Berkeley: University of California Press.

Ramirez, Susan Elisabeth. 1996. *The World Upside Down: Cross-Cultural Contact and Conflict in Sixteenth-Century Peru.* Stanford: Stanford University Press.

Sallnow, Michael. 1987. *Pilgrims of the Andes: Regional Cults in Cusco.* Washington, DC: Smithsonian Institution Press.

Seligmann, Linda and Kathleen Fine-Dare, eds. 2019. *The Andean World.* Abingdon and New York: Routledge.

Stensrud, Astrid. 2021. *Watershed Politics and Climate Change in Peru.* London: Pluto Press.

Stern, Steve. 1983. *Peru's Indian Peoples and the Challenge of Spanish Conquest, Huamanga to 1640.* Madison: University of Wisconsin Press.

Urton, Garry. 1981. *At the Crossroads of the Earth and the Sky: An Andean Cosmology.* Austin: University of Texas Press.

van den Berghe, Pierre and George Primov. 1977. *Inequality in the Peruvian Andes: Class and Ethnicity on Cuzco.* Columbia: University of Missouri Press.

Index

Adi 159, 167–8
affective geomorphology 78
agency 4, 7, 13–5, 23, 48, 50, 52, 64, 88, 90, 92, 100–1, 124, 138–9, 141, 158, 160, 167, 176, 179, 185, 194, 198, 206, 219
Agents 77, 217–9
Alaknanda 15, 138, 151
Alfthan, Björn 3
Allison, Elizabeth 12, 77, 91, 177
Alter, Stephen 20
Ama Jomo 85, 179–80, 182–93
amplification 1
Amrith and Smyer Yü 16, 19, 124, 201
Amrith, Sunil 8, 16, 19, 124, 161, 201
Andersen, Oberborbeck 1, 27
Andes 6, 14, 87, 100, 215–21
Animals 11, 23, 32, 39–40, 43, 53–5, 72, 76–7, 107, 109, 112, 114, 126–8, 143, 182–5, 200, 208, 215, 217–8, 221
anthropocenic water 24, 48, 50
anthropogenic 1, 2, 12, 15–24, 32, 35, 49–52, 54, 64–5, 88, 100, 111, 123, 125–6, 128, 131–5, 139–40, 142, 144, 157–8, 160, 164–5, 169, 180, 193, 218, 220
anthropogenic climate change 1, 3, 7, 11, 20, 22, 64, 191
anthropogenic thirst 126, 132, 135
anthropogenic transformations 2, 11, 20, 31–2, 40, 42, 49, 85, 132, 164–5, 193
aquatic land 18, 50–1, 218
Architectural structures 140, 144, 152
Assam 16, 73, 106–8, 110–2, 119, 129, 131, 155, 160–1, 163, 216
atmosphere 19, 52, 139, 141, 150, 152–3

Banerjee and Wouters 12, 41, 74, 175
Bangladesh 9, 16, 18, 23, 157–8, 161–3, 197–8, 200–3, 205, 207, 209–10
Baptized 1
Barad, Karen 10
basket 37, 39, 145–8
Bay of Bengal 16, 111, 157, 161, 201
Bhagirathi 15, 138, 151
Bhutia, Kalzang Dorjee 5, 125–6
Bonbibi 203, 205
border 17, 22–4, 30, 49, 73, 157, 159–60, 162–70, 180, 197, 202, 205, 207, 209, 210, 219, 221
Brahmaputra 16, 23–4, 34, 107–8, 110–3, 117–8, 157–70, 218, 221
Budyko, Mikhail 1
Budyko–Sellers Climate Model 1

cardamom 124, 130–1
Cenozoic 14
Chair 145, 151–2, 154
Chakraborty and Sherpa 5, 91
cham 57
Chao and Celermajer 7
Chao and Enari 4, 78
Chao, Sophie 4, 7, 78
Char Dham Yatra 138, 141, 144, 147, 151, 153–4
China 18, 22–3, 33–4, 65, 113, 130, 157–60, 162–70, 182, 217, 221
Chomolungma 22, 48, 87, 89
Climate 2–13, 15–20, 22–4, 31–3, 41–2, 49–52, 60–1, 65, 71–3, 84–5, 88, 91, 100, 106–9, 123–32, 134–5, 139–40, 142, 157–8, 160, 169, 174–5, 177–8, 180, 182–5, 193–4, 198–202, 206, 209, 211, 216, 219
climate change 1–13, 20, 22, 24, 32–3, 41–3, 60, 64–5, 72, 84–5, 87–8, 91, 93, 100, 102, 123, 125, 132, 134–5, 144, 157, 159, 174–6, 179, 185, 191–5, 197–9, 202, 204, 209–11, 215–6, 219–20, 222

clime 1, 5–13, 15–8, 20, 23–4, 30–3, 37–8, 42–3, 49–50, 52–3, 58, 61, 63, 71–3, 75–8, 81, 84–5, 87, 90, 100, 106–15, 117–9, 123–35, 139–41, 144–5, 148–50, 152–4, 157–67, 169–70, 174–9, 185, 192–4, 197–202, 206–8, 210–11
clime-change 8, 10, 31–2, 43, 63, 71, 84–5, 192
clime-studies 5–6, 8–12, 15, 20, 32–3, 49, 72, 77, 90, 107, 123, 158, 174
clime-thinking 6, 106, 140, 148–9
climing 6, 12, 15, 49–50, 52, 58, 61–2, 87–9, 101, 106–7, 109–10, 112, 116, 118, 138, 142, 145, 150, 151, 175, 185, 193
Clouse, Carey 21
colonial rain clime 123, 128, 130–2, 135
communication 11, 52, 57, 79, 93, 99, 107, 112, 175–6, 220, 222
communities of life 8–9, 11, 18, 49, 158, 169, 174, 177
comparison 8, 216
Cons, Jason 5, 11, 16, 210, 218
conservation 12, 77, 98, 162, 197–8, 202–4, 209–11
cosmology 39, 41, 57, 62, 77, 203
cosmovision 64, 183
cotton 118, 130
cows 128, 206–7
Critical Zone 49, 218
crops 11, 16, 32, 35–6, 40, 43, 109, 112, 124, 130, 217
Crosby, Alfred W. 2, 25, 217
cultivation 32, 34–6, 43, 112, 130–1

D'Avignon, Robyn 7, 15, 79, 178, 193
Darjeeling Rain Clime 124
Davis, Alexander 23, 221
Deep Time 4, 13–4, 23, 31, 48, 76–7, 119, 124–5, 138, 145, 151, 158, 161, 165, 169, 177, 185, 187, 220–1
Delley and Aiyadurai 5
delta 11, 16, 112, 152, 160, 197, 199–204, 206–8, 211, 215
dhu 12, 181–2
Dibang 168
digital media 101
Dodds and Nutall 1
Dorji, Kinley 13, 56, 65, 68, 218–19
drib 53–4, 57–60, 62–3, 65, 177–8, 182, 190
dry-rice 34
Dumka, Umesh Chandra 20
dunkhar 53–4, 56, 61

earth architecture 4, 7, 14–5, 22, 138–9, 144
ecological services 15, 184
ecosublime 144
environmental science 71
Eriksen, Thomas Hyland 2, 92
Everest 87–91, 93–8, 110–2, 220

Fleetwood, Lachlan 14, 87
Fleming, James Rodger 5, 32
food 11, 30, 32, 37–8, 42, 111–2, 185–9, 206, 218

Gagné, Karine 5, 7, 12, 21, 92, 176
Gamukh 143, 148, 152
Ganges 15, 152, 218, 220
Garhwal Himalayas 15, 138–9, 220
geopolitical clime 23, 158, 160, 162, 164–6, 169
glaciers 2–4, 7, 12, 14, 18–21, 50, 52–4, 64, 77, 92–3, 100, 124, 176, 180–1, 184, 194, 215–7, 219–20, 222
globalization 95, 222
guidebooks 139
guides 88–9, 91–2, 96, 98–100, 183, 220
Guru Rinpoche 56, 74, 80–1, 85, 176

Haberman, David L. 100, 175
Haraway, Donna 8
Heneise, Michael T. 39, 62, 167
Himalaya 2–9, 11–24, 34, 40, 48, 51–2, 64–5, 82, 85, 87, 89, 91, 99, 123–8, 130, 132, 135, 138–40, 142, 144–5, 149, 157–8, 160–2, 165–6, 174, 176–80, 201, 215–221
Himalayan multispecies climes 5
History 7, 9, 13, 15, 20, 22, 30–3, 37, 48–50, 71, 75–6, 79–81, 84, 88, 90, 98–9, 101, 107–8, 117, 119, 123, 125, 135, 139, 159, 161–2, 165, 176–7, 194, 198, 201–3, 209, 218, 221–2
holy water 56, 187
Huber, Toni 12, 34, 53–4, 60, 167, 177, 193–4
Hulme, Mike 7–8, 107
hydro-humanities 49

ice 1, 7, 12, 19, 21–3, 51, 87–94, 96, 99–102, 108–9, 157, 161–4, 181, 216, 217
ice-stupas 21–3
images 22, 41, 87–90, 96–8, 100–1, 202, 209, 219
imagination 8, 10, 19, 49, 76, 78, 87, 89, 95, 101, 112, 113, 115, 116, 140–1, 151, 175, 179, 192, 198–200, 202–4, 211, 215

India 9, 11, 17–8, 22–3, 30–1, 33–5, 40, 48, 81, 89, 110–1, 113–4, 124, 128, 138, 143, 157–70,182, 197, 202–3, 205, 21, 217, 221
indigeneity 30, 39–41
indigenous 4–5, 7, 9–11, 13, 20, 23, 30–4, 37, 39–43, 48, 71–2, 92, 94, 109, 123, 125–7, 130–5, 157–8, 164, 168–9, 193–4, 217, 221
indigenous rain clime 123, 125–7, 130–5
indigenous rice 11, 30–2, 43
infrastructure 22, 87–8, 91, 99–100, 102, 133–5, 138, 140, 144, 148, 158–62, 164–5, 170
Ingold, Tim 6, 9, 142, 145–7, 149–51, 153, 178
Irvine, Richard D.G 4, 48, 109, 139–41, 151, 153

Jalais, Annu 12, 199
jhum 34–5
jigdra 61–3, 67–8
Jomolhari 80, 82, 180–5
jungwa zhi 56, 191

Kimmerer, Robin Wall 71–2
Kire, Easterine 37
Klaver, Irene J. 7, 160
Knox, Hannah 4, 8
Kohn, Eduardo 7
Kumari, Nikhul 20

lake 7, 21–3, 48–64, 66–8, 76, 79, 85, 126–7, 133, 163, 187, 189–92, 215, 217–9
lake-yaks 53–4, 58
landscape 4, 10–1, 13–4, 22, 32, 35, 41, 51, 53, 54, 65, 71, 73–81, 83–5, 93, 96, 99, 108, 110–11, 113–4, 123, 125, 130, 133, 135, 139–40, 142–4, 152, 164, 176–7, 183, 185–7, 191, 194, 199, 201–2, 204, 207, 215, 218–20
Langenbrunner, Baird 3
Lepcha 126–7, 131
lhakhang 79–82, 189
lively ethography 7, 11
local rice 30

Mandakini 15, 138, 151
Mathur, Nayanika 5, 200, 211
McNeill and Engelke 2, 27
Media 93, 98, 100–1, 164
Medog County 166
Mermaids 21, 48, 50, 55, 218
migration 33–4, 64, 125, 132, 134, 162, 221–2
mobile phones 99, 101
modern water 50
monsoon 7, 16, 19–20, 43, 106–19, 124, 127–8, 131, 161, 211
monsoon clime 20, 107–8, 110–13, 115, 118
more-than-lakes 22
Morton, Timothy 10
Mount Kanchenjunga 126, 160
mountain clime 12, 139, 174–9, 185, 192
mountains 7, 12–8, 20–1, 31, 52, 54–5, 74, 76, 77, 80, 82–4, 88, 91–3, 96–9, 101, 115, 126, 129–30, 138–42, 145, 150–51, 154, 158–61, 163, 165–7, 174–9, 183, 185, 191, 193–5, 201, 215–7, 219–21
Multipolar 7–8, 18–9, 87, 90
multipolar climes 87
multispecies 1, 4–18, 20, 31–3, 35, 38–40, 49, 72–3, 75–8, 84–5, 107, 112, 123, 143, 158–62, 164–6, 169, 174, 200, 221
Murray, Ainslie 14, 138, 220

Naga 11, 30–43, 56, 74, 124, 127, 217
Naga Hills 31, 34, 134
Naga tribe 33–4, 36–7
Naga uplands 11, 30–2, 34–6, 41–3
Nagaland 30–1, 34, 42–3
Namcha Barwa 166–7
namshey jurwa 185, 191
Nepal 9, 22, 87–9, 91, 94, 96–9, 124, 131, 215, 220, 222
Nepalis 97, 131
ngodrup 53, 56, 58–9, 61–2
nonhuman 40, 52, 75, 82, 88, 100, 107, 109–12, 114, 118, 123, 175, 184, 194, 197, 200, 205–6, 208, 211, 216–9
nonsecular beings 175, 185, 195
northeast India 18, 33–4, 161–2, 165

Other-than-humans 5–6, 10, 39, 41, 49, 52, 55, 73, 75–7, 107, 109, 135, 169, 176–7

paddy 7, 11, 30–43, 111–2, 207
Paerregaard, Karsten 6–7, 49, 87, 106, 157, 219, 221
Pandit, Maharaj K. 14–5, 19–20, 50
Parvaiz, Athar 21
Pemakö 166
phodrang 185–6, 188–92, 194
pilgrimage 15, 74, 79, 95, 138–43, 148–50, 153–4, 161, 220
pilgrims 14–5, 74, 138–46, 148–9, 153, 220–1
plantation 7, 43, 129–32

plants 7, 11, 16, 31, 36, 38, 40, 55, 76–7, 107, 111, 126–8, 130–1, 143, 199, 215, 217, 221
polar amplification 1
politics 4, 33, 42, 132, 158–9, 163, 200, 209, 211
porters 88–9, 91–2, 98, 100, 133
Povinelli, Elizabeth 7
predation 12, 198, 200, 204, 208
pre-modern Brahmaputra Valley 16

rain 7, 16, 20, 32–3, 42–3, 53, 58, 60, 64, 67–8, 106, 108, 110, 112–3, 115–8, 123–35, 161, 164, 186, 191, 201, 216–7
re-climing 6, 158, 164, 169
relational Himalayas 13
relationality 5–6, 12, 15–6, 52, 64, 77, 165
religious tourism 143
rice 11, 30–9, 41–2, 43, 110–2, 118, 127, 130–1, 207, 217
rice beer 39
rice blast fungus 43
ritual 9, 12–3, 15, 21, 32, 37–9, 42, 50, 56–8, 61, 64–5, 78–9, 81, 88, 107, 117, 126, 132, 134–5, 142, 175–7, 179, 182–3, 185–8, 192–5, 218–9
ritual geology 79, 178, 193–4
ritual specialist 127
river sources 138, 140–2
rivers 14–6, 18, 41, 48, 50–1, 55–6, 75–7, 82, 91, 110, 112, 124–27, 138, 142, 149–51, 154, 158, 160, 163, 168, 170, 177, 201–2, 211, 215–20

Saikia, Arupjyoti 16, 34, 107, 110–1, 160, 163, 165
samaj 134
Schendel, Willem van 16, 201
science-making 2, 4, 7
Searle, Mike 7, 14, 16–17, 20
Sellers, Willem 1
sentinel 12, 198, 209
Serreze, Mark 1, 5, 27–8
Shack 148–9, 151–2, 154
shrimp 202, 208, 210
Shroder, John F. 20
Siang 166–9
Sinanan, Jolynna 22, 87, 220
Smyer Yü and Wouters 5, 8, 19, 31–2, 90, 123, 174
Smyer Yü, Dan 2, 5–8, 12–3, 16, 18–9, 31–2, 40, 48–9, 51, 72, 87–8, 90–1, 93, 123–5, 144, 157, 160, 174, 178, 198, 201
Snively and Corsiglia 9
social media 88–9, 97–8, 102
Sörlin, Sverker 22, 50, 90
Southern Naga Hills 30, 217
Soviet Union 1
spiritual 7, 10–3, 15, 21, 23, 30–3, 35, 37, 39, 49–50, 53, 55–8, 60, 71–9, 81, 83, 85, 87–8, 91, 93, 95, 107, 113, 123, 126, 132–3, 138, 142, 143, 151, 153, 175, 178, 180, 184–5, 187, 191, 193–4, 215, 218, 220
springs 125–7, 132–5, 217, 220
Srivastava, Ankur 20
Sturm, Perovich, and Serreze 1, 28
Sundaram and Holland 18
Sundarbans 11–2, 16, 18, 197–211, 218
superfine rice 30

Tamang, Sangay 20, 123, 217
tea 33, 106, 124, 129–31, 149
Tendong Mountain 126
Tent 148, 151, 152, 154
Tenyimi 33, 35, 37–8, 42–3
terma 56, 61
terrestrial 1, 5–6, 8–11, 15–20, 32–3, 48, 51, 57, 60, 74, 78, 80, 84–5, 107–9, 123, 139, 141, 145, 158, 160, 174–5, 177, 179, 182, 194, 218
the four elements 191
Third Pole 18
Tibetan Plateau 17–9, 111, 178
tiger 7, 11–2, 112, 162, 166, 197–200, 202–11, 218
toponyms 7, 71, 73–6, 78, 81, 83–4, 218
tourism 87–9, 91–2, 96–7, 99–101, 143–4, 148, 166, 184, 220–22
trekking 96–9, 141, 220
trekking poles 144
tshechu 57
tshomen 21–2, 50, 53, 55–68, 89, 218
tshomen-lakes 21, 50, 55, 58, 63–4, 66–7
Tshomjay 62
Tsosie and Claw 9
Tuting 168

Upper Siang Project 166–7, 169
US Apollo project 1

Van Dooren and Bird Rose 7, 90

Walsh and Rackauckas 1, 28
water 3–4, 7, 14–7, 19–21, 23–4, 31–2, 35–6, 48–53, 56–7, 62–8, 80, 89–91, 99–100, 106–7, 110–14, 117–8, 123–135, 140, 150, 152, 157–8, 160–170, 176, 184–5, 187, 191, 197, 200–1, 206–7, 215–21

weather 4–6, 9, 10, 12, 14, 19, 21, 23, 31, 35, 37, 42–3, 57–61, 65, 67, 87, 93–4, 96, 106–9, 115, 117–8, 127, 129, 131, 144–5, 157, 175, 182–7, 191, 200, 206, 208, 215–17
weather-worlds 6, 21
Wegener, Alfred 14
wet-rice 34–5
Wijunamai, Roderick 11, 41, 217
wind 6–7, 16, 19, 21, 38, 60, 106, 109, 111, 115, 148, 161, 176, 182–83, 185, 191, 216
work 8, 14, 22, 35–7, 42, 76, 91–2, 96, 98–9, 107–8, 114–17, 131, 134, 143–44, 186, 192, 197–200, 205–6, 210–11
Wouters and Heneise 15, 167
Wouters, Jelle J.P. 1–2, 4–6, 8, 12, 15, 19, 21, 31–2, 39, 41, 48, 53, 55, 72, 74, 80, 84, 90, 123, 139–41, 167, 174–5, 177, 179–80, 216, 218, 219

Yamuna 15, 138, 151
Yangzom, Deki 12, 84–5, 139, 179, 185, 188–90, 192, 194, 219
Yingkiong 168–69
yul lha 12, 55, 177, 179, 183, 193, 195
Yusoff, Kathryn 13
zamling kabenob 13, 185, 195

Zee, Jerry C. 7–8, 17

For Product Safety Concerns and Information please contact our EU
representative GPSR@taylorandfrancis.com
Taylor & Francis Verlag GmbH, Kaufingerstraße 24, 80331 München, Germany

www.ingramcontent.com/pod-product-compliance
Lightning Source LLC
Chambersburg PA
CBHW060253220326
41598CB00027B/4082

* 9 7 8 1 0 3 2 7 7 7 0 0 9 *